High-energy Neutrino Astronomy: The Cosmic Ray Connection

Francis Halzen and Dan Hooper

Abstract

This is a review of neutrino astronomy anchored to the observational fact that Nature accelerates protons and photons to energies in excess of 10^{20} and 10^{13} eV, respectively.

Although the discovery of cosmic rays dates back close to a century, we do not know how and where they are accelerated. There is evidence that the highest energy cosmic rays are extra-galactic — they cannot be contained by our galaxy's magnetic field anyway because their gyroradius far exceeds its dimension. Elementary elementary-particle physics dictates a universal upper limit on their energy of 5×10^{19} eV, the so-called Greisen-Kuzmin-Zatsepin cutoff; however, particles in excess of this energy have been observed by all experiments, adding one more puzzle to the cosmic ray mystery. Mystery is fertile ground for progress: we will review the facts as well as the speculations about the sources.

There is a realistic hope that the oldest problem in astronomy will be resolved soon by ambitious experimentation: air shower arrays of $10^4 \, \text{km}^2$ area, arrays of air Cerenkov detectors and, the subject of this review, kilometer-scale neutrino observatories.

We will review why cosmic accelerators are also expected to be cosmic beam dumps producing associated high-energy photon and neutrino beams. We will work in detail through an example of a cosmic beam dump, gamma ray bursts. These are expected to produce neutrinos from MeV to EeV energy by a variety of mechanisms. We will also discuss active galaxies and GUT-scale remnants, two other classes of sources speculated to be associated with the highest energy cosmic rays. Gamma ray bursts and active galaxies are also the sources of the highest energy gamma rays, with emission observed up to 20 TeV, possibly higher.

The important conclusion is that, independently of the specific blueprint of the source, it takes a kilometer-scale neutrino observatory to detect the neutrino beam associated with the highest energy cosmic rays and gamma rays. We also briefly review the ongoing efforts to commission such instrumentation.

Contents

I. The Highest Energy Particles: Cosmic Rays, Photons and Neutrinos   4
    A. The New Astronomy   4
    B. The Highest Energy Cosmic Rays: Facts   6
    C. The Highest Energy Cosmic Rays: Fancy   8
       1. Acceleration to > 100 EeV?   8
       2. Are Cosmic Rays Really Protons: the GZK Cutoff?   10
       3. Could Cosmic Rays be Photons or Neutrinos?   11
    D. A Three Prong Assault on the Cosmic Ray Puzzle   13
       1. Giant Cosmic Ray Detectors   13
       2. Gamma rays from Cosmic Accelerators   14
       3. Neutrinos from Cosmic Accelerators   17

II. High-energy Neutrino Telescopes   19
    A. Observing High-energy Neutrinos   19
    B. Large Natural Cerenkov Detectors   22
       1. Baikal, ANTARES, Nestor and NEMO: Northern Water   25
       2. AMANDA: Southern Ice   28
       3. IceCube: A Kilometer-Scale Neutrino Observatory   33
    C. EeV Neutrino Astronomy   35

III. Cosmic Neutrino Sources   37
    A. A List of Cosmic Neutrino Sources   37
    B. Gamma Ray Bursts: A Detailed Example of a Generic Beam Dump   39
       1. GRB Characteristics   39
       2. A Brief History of Gamma Ray Bursts   40
       3. GRB Progenitors?   41
       4. Fireball Dynamics   42
       5. Ultra High-energy Protons From GRB?   47
       6. Neutrino Production in GRB: the Many Opportunities   49
       7. Thermal MeV Neutrinos from GRB   50
       8. Shocked Protons: PeV Neutrinos   51

- 9. Stellar Core Collapse: Early TeV Neutrinos — 53
- 10. UHE Protons From GRB: EeV Neutrinos — 55
- 11. The Decoupling of Neutrons: GeV Neutrinos — 57
- 12. Burst-To-Burst Fluctuations and Neutrino Event Rates — 59
- 13. The Effect of Neutrino Oscillations — 61
- C. Blazars: the Sources of the Highest Energy Gamma rays — 62
  - 1. Blazar Characteristics — 62
  - 2. Blazar Models — 63
  - 3. Highly Shocked Protons: EeV Blazar Neutrinos — 64
  - 4. Moderately Shocked Protons: TeV Blazar Neutrinos — 66
- D. Neutrinos Associated With Cosmic Rays of Top-Down Origin — 67
  - 1. Nucleons in Top-Down Scenarios — 68
  - 2. Neutrinos in Top-Down Scenarios — 69

IV. The Future for High-energy Neutrino Astronomy — 71

Acknowledgments — 71

References — 71

I. THE HIGHEST ENERGY PARTICLES: COSMIC RAYS, PHOTONS AND NEUTRINOS

A. The New Astronomy

Conventional astronomy spans 60 octaves in photon frequency, from 10^4 cm radio-waves to 10^{-14} cm gamma rays of GeV energy; see Fig. 1. This is an amazing expansion of the power of our eyes which scan the sky over less than a single octave just above 10^{-5} cm wavelength. This new astronomy probes the Universe with new wavelengths, smaller than 10^{-14} cm, or photon energies larger than 10 GeV. Besides the traditional signals of astronomy, gamma rays, gravitational waves, neutrinos and very high-energy protons become astronomical messengers from the Universe. As exemplified time and again, the development of novel ways of looking into space invariably results in the discovery of unanticipated phenomena. As is the case with new accelerators, observing only the predicted will be slightly disappointing.

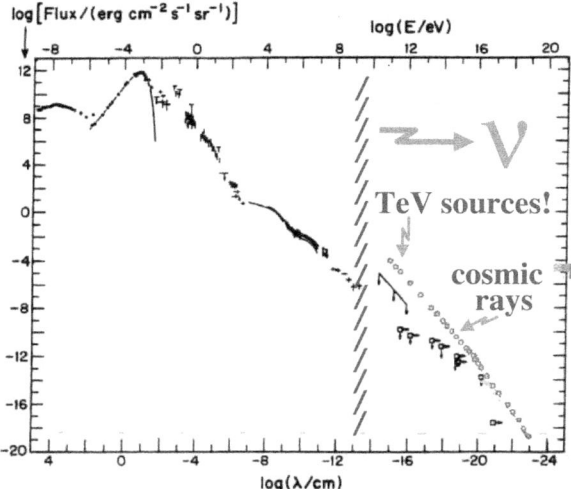

FIG. 1: The diffuse flux of photons in the Universe, from radio waves to GeV-photons. Above tens of GeV, only limits are reported although individual sources emitting TeV gamma rays have been identified. Above GeV energy, cosmic rays dominate the spectrum.

Why pursue high-energy astronomy with neutrinos or protons despite considerable instrumental challenges? A mundane reason is that the Universe is not transparent to photons of TeV energy and above (units are: GeV/TeV/PeV/EeV/ZeV in ascending factors of 10^3). For instance, a PeV energy photon cannot deliver information from a source at the edge of our own galaxy because it will annihilate into an electron pair in an encounter with a 2.7 Kelvin microwave photon before reaching our telescope. In general, energetic photons are absorbed on background light by pair production $\gamma + \gamma_{\rm bkgnd} \to e^+ + e^-$ of electrons above a threshold E given by

$$4E\epsilon \sim (2m_e)^2, \tag{1}$$

where E and ϵ are the energy of the high-energy and background photon, respectively. Eq. (1) implies that TeV-photons are absorbed on infrared light, PeV photons on the cosmic microwave background and EeV photons on radio-waves; see Fig. 1. Only neutrinos can reach us without attenuation from the edge of the Universe.

At EeV energies, proton astronomy may be possible. Near 50 EeV and above, the arrival directions of electrically charged cosmic rays are no longer scrambled by the ambient magnetic field of our own galaxy. They point back to their sources with an accuracy determined by their gyroradius in the intergalactic magnetic field B:

$$\theta \cong \frac{d}{R_{\rm gyro}} = \frac{dB}{E}, \tag{2}$$

where d is the distance to the source. Scaled to units relevant to the problem,

$$\frac{\theta}{0.1°} \cong \frac{\left(\frac{d}{1 \text{ Mpc}}\right)\left(\frac{B}{10^{-9}\text{ G}}\right)}{\left(\frac{E}{3\times 10^{20}\text{ eV}}\right)}. \tag{3}$$

Speculations on the strength of the inter-galactic magnetic field range from 10^{-7} to 10^{-12} Gauss in the local cluster. For a distance of 100 Mpc, the resolution may therefore be anywhere from sub-degree to nonexistent. It is still possible that the arrival directions of the highest energy cosmic rays provide information on the location of their sources. Proton astronomy should be possible; it may also provide indirect information on intergalactic magnetic fields. Determining the strength of intergalactic magnetic fields by conventional astronomical means has been challenging.

B. The Highest Energy Cosmic Rays: Facts

In October 1991, the Fly's Eye cosmic ray detector recorded an event of energy $3.0 \pm^{0.36}_{0.54} \times 10^{20}$ eV [1]. This event, together with an event recorded by the Yakutsk air shower array in May 1989 [2], of estimated energy $\sim 2 \times 10^{20}$ eV, constituted (at the time) the two highest energy cosmic rays ever seen. Their energy corresponds to a center of mass energy of the order of 700 TeV or $\sim$ 50 Joules, almost 50 times the energy of the Large Hadron Collider (LHC). In fact, all active experiments [3] have detected cosmic rays in the vicinity of 100 EeV since their initial discovery by the Haverah Park air shower array [4]. The AGASA air shower array in Japan[5] has now accumulated an impressive 10 events with energy in excess of 10^{20} eV [6].

The accuracy of the energy resolution of these experiments is a critical issue. With a particle flux of order 1 event per km^2 per century, these events are studied by using the earth's atmosphere as a particle detector. The experimental signature of an extremely high-energy cosmic particle is a shower initiated by the particle. The primary particle creates an electromagnetic and hadronic cascade. The electromagnetic shower grows to a shower maximum, and is subsequently absorbed by the atmosphere.

The shower can be observed by: i) sampling the electromagnetic and hadronic components when they reach the ground with an array of particle detectors such as scintillators, ii) detecting the fluorescent light emitted by atmospheric nitrogen excited by the passage of the shower particles, iii) detecting the Cerenkov light emitted by the large number of particles at shower maximum, and iv) detecting muons and neutrinos underground.

The bottom line on energy measurement is that, at this time, several experiments using the first two techniques agree on the energy of EeV-showers within a typical resolution of 25%. Additionally, there is a systematic error of order 10% associated with the modeling of the showers. All techniques are indeed subject to the ambiguity of particle simulations that involve physics beyond the LHC. If the final outcome turns out to be an erroneous inference of the energy of the shower because of new physics associated with particle interactions at the Λ_{QCD} scale, we will be happy to contemplate this discovery instead.

Could the error in the energy measurement be significantly larger than 25%? The answer is almost certainly negative. A variety of techniques have been developed to overcome the fact that conventional air shower arrays do calorimetry by sampling at a single depth. They

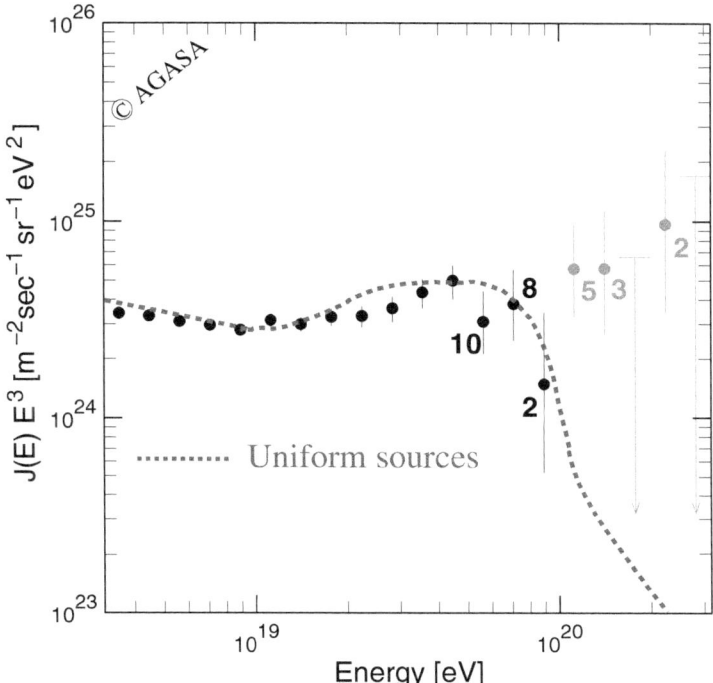

FIG. 2: The cosmic ray spectrum peaks in the vicinity of 1 GeV and has features near 10^{15} and 10^{19} eV referred to as the "knee" and "ankle" in the spectrum, respectively. Shown is the flux of the highest energy cosmic rays near and beyond the ankle measured by the AGASA experiment. Note that the flux is multiplied by E^3.

also give results within the range already mentioned. So do the fluorescence experiments that embody continuous sampling calorimetry. The latter are subject to understanding the transmission of fluorescent light in the dark night atmosphere — a challenging problem given its variation with weather. Stereo fluorescence detectors will eventually eliminate this last hurdle by doing two redundant measurements of the same shower from different locations. The HiRes collaborators have one year of data on tape which should allow them to settle energy calibration once and for all.

The premier experiments, HiRes and AGASA, agree that cosmic rays with energy in excess of 10 EeV are not galactic in origin and that their spectrum extends beyond 100 EeV.

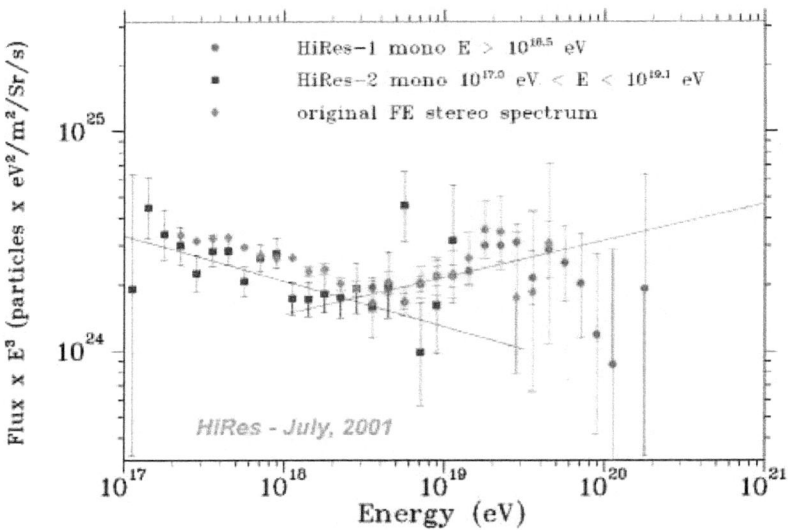

FIG. 3: As in Fig. 2, but as measured by the HiRes experiment.

They disagree on almost everything else. The AGASA experiment claims evidence that the highest energy cosmic rays come from point sources, and that they are mostly heavy nuclei. The HiRes data do not support this. Because of such low statistics, interpreting the measured fluxes as a function of energy is like reading tea leaves; one cannot help however reading different messages in the spectra (see Fig. 2 and Fig. 3).

C. The Highest Energy Cosmic Rays: Fancy

1. Acceleration to > 100 EeV?

It is sensible to assume that, in order to accelerate a proton to energy E in a magnetic field B, the size R of the accelerator must be larger than the gyroradius of the particle:

$$R > R_{\text{gyro}} = \frac{E}{B}. \tag{4}$$

That is, the accelerating magnetic field must contain the particle orbit. This condition yields a maximum energy

$$E = \gamma B R \tag{5}$$

TABLE I: Requirements to generate the highest energy cosmic rays in astrophysical sources.

	Conditions with $E \sim 10$ EeV		
• Quasars	$\gamma \cong 1$	$B \cong 10^3$ G	$M \cong 10^9 M_{\text{sun}}$
• Blazars	$\gamma \gtrsim 10$	$B \cong 10^3$ G	$M \cong 10^9 M_{\text{sun}}$
• Neutron Stars	$\gamma \cong 1$	$B \cong 10^{12}$ G	$M \cong M_{\text{sun}}$
Black Holes			
$\vdots$			
• GRB	$\gamma \gtrsim 10^2$	$B \cong 10^{12}$ G	$M \cong M_{\text{sun}}$

by dimensional analysis and nothing more. The γ-factor has been included to allow for the possibility that we may not be at rest in the frame of the cosmic accelerator. The result would be the observation of boosted particle energies. Theorists' imagination regarding the accelerators has been limited to dense regions where exceptional gravitational forces create relativistic particle flows: the dense cores of exploding stars, inflows on supermassive black holes at the centers of active galaxies, annihilating black holes or neutron stars. All speculations involve collapsed objects and we can therefore replace R by the Schwartzschild radius

$$R \sim GM/c^2 \qquad (6)$$

to obtain

$$E \propto \gamma BM. \qquad (7)$$

Given the microgauss magnetic field of our galaxy, no structures are large or massive enough to reach the energies of the highest energy cosmic rays. Dimensional analysis therefore limits their sources to extragalactic objects; a few common speculations are listed in Table 1.

Nearby active galactic nuclei, distant by ~ 100 Mpc and powered by a billion solar mass black holes, are candidates. With kilogauss fields, we reach 100 EeV. The jets (blazars) emitted by the central black hole could reach similar energies in accelerating substructures (blobs) boosted in our direction by Lorentz factors of 10 or possibly higher. The neutron star or black hole remnant of a collapsing supermassive star could support magnetic fields of 10^{12} Gauss, possibly larger. Highly relativistic shocks with $\gamma > 10^2$ emanating from the collapsed black hole could be the origin of gamma ray bursts and, possibly, the source of the highest energy cosmic rays.

The above speculations are reinforced by the fact that the sources listed are also the sources of the highest energy gamma rays observed. At this point, however, a reality check is in order. The above dimensional analysis applies to the Fermilab accelerator: 10 kilogauss fields over several kilometers corresponds to 1 TeV. The argument holds because, with optimized design and perfect alignment of magnets, the accelerator reaches efficiencies matching the dimensional limit. It is highly questionable that nature can achieve this feat. Theorists can imagine acceleration in shocks with an efficiency of perhaps 10%.

The astrophysics problem of obtaining such high-energy particles is so daunting that many believe that cosmic rays are not the beams of cosmic accelerators but the decay products of remnants from the early Universe, such as topological defects associated with a Grand Unified Theory (GUT) phase transition.

2. *Are Cosmic Rays Really Protons: the GZK Cutoff?*

All experimental signatures agree on the particle nature of the cosmic rays — they look like protons or, possibly, nuclei. We mentioned at the beginning of this article that the Universe is opaque to photons with energy in excess of tens of TeV because they annihilate into electron pairs in interactions with the cosmic microwave background. Protons also interact with background light, predominantly by photoproduction of the Δ-resonance, i.e. $p + \gamma_{CMB} \to \Delta \to \pi + p$ above a threshold energy E_p of about 50 EeV given by:

$$2E_p \epsilon > \left(m_\Delta^2 - m_p^2\right). \quad (8)$$

The major source of proton energy loss is photoproduction of pions on a target of cosmic microwave photons of energy ϵ. The Universe is, therefore, also opaque to the highest energy cosmic rays, with an absorption length of

$$\lambda_{\gamma p} = \left(n_{\text{CMB}}\, \sigma_{p+\gamma_{\text{CMB}}}\right)^{-1} \quad (9)$$

$$\cong 10 \text{Mpc}, \quad (10)$$

when their energy exceeds 50 EeV. This so-called GZK cutoff establishes a universal upper limit on the energy of the cosmic rays. The cutoff is robust, depending only on two known numbers: $n_{\text{CMB}} = 400\,\text{cm}^{-3}$ and $\sigma_{p+\gamma_{\text{CMB}}} = 10^{-28}\,\text{cm}^2$ [8, 9, 10, 11].

Protons with energy in excess of 100 EeV, emitted in distant quasars and gamma ray bursts, will lose their energy to pions before reaching our detectors. They have, nevertheless, been observed, as we have previously discussed. They do not point to any sources within the GZK-horizon however, i.e. to sources in our local cluster of galaxies. There are three possible resolutions: i) the protons are accelerated in nearby sources, ii) they do reach us from distant sources which accelerate them to even higher energies than we observe, thus exacerbating the acceleration problem, or iii) the highest energy cosmic rays are not protons.

The first possibility raises the challenge of finding an appropriate accelerator by confining these already unimaginable sources to our local galactic cluster. It is not impossible that all cosmic rays are produced by the active galaxy M87, or by a nearby gamma ray burst which exploded a few hundred years ago.

Stecker [12] has speculated that the highest energy cosmic rays are Fe nuclei with a delayed GZK cutoff. The details are complicated but the relevant quantity in the problem is $\gamma = E/AM$, where A is the atomic number and M the nucleon mass. For a fixed observed energy, the smallest boost above GZK threshold is associated with the largest atomic mass, i.e. Fe.

3. Could Cosmic Rays be Photons or Neutrinos?

Topological defects predict that the highest energy cosmic rays are predominantly photons. A topological defect will suffer a chain decay into GUT particles X and Y, that subsequently decay to familiar weak bosons, leptons and quark or gluon jets. Cosmic rays are, therefore, predominately the fragmentation products of these jets. We know from accelerator studies that, among the fragmentation products of jets, neutral pions (decaying into photons) dominate, in number, protons by close to two orders of magnitude. Therefore, if the decay of topological defects is the source of the highest energy cosmic rays, they must be photons. This is a problem because there is compelling evidence that the highest energy cosmic rays are not photons:

1. The highest energy event observed by Fly's Eye is not likely to be a photon [7]. A photon of 300 EeV will interact with the magnetic field of the earth far above the atmosphere and disintegrate into lower energy cascades — roughly ten at this particular energy. The detector subsequently collects light produced by the fluorescence of atmospheric nitrogen along the path of the high-energy showers traversing the atmosphere. The anticipated shower profile of a 300 EeV photon is shown in Fig. 4. It disagrees with the data. The

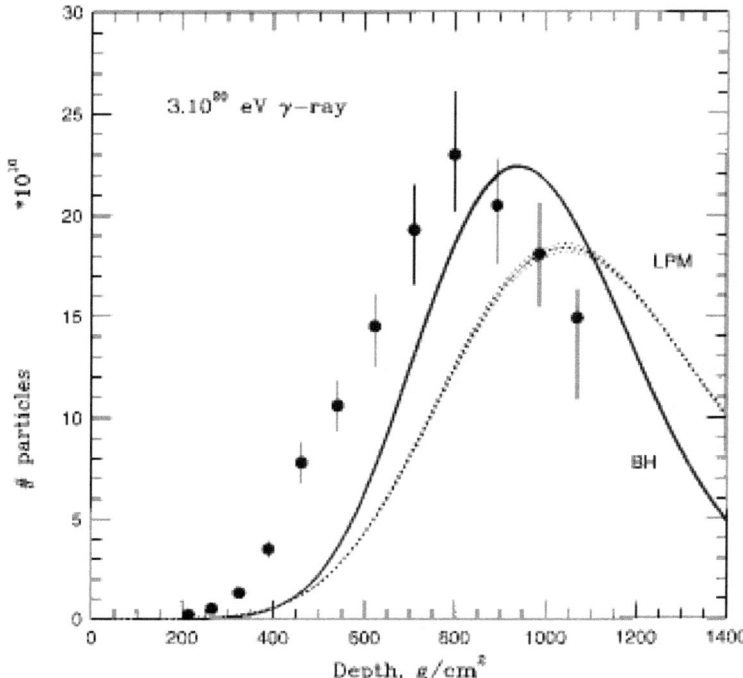

FIG. 4: The composite atmospheric shower profile of a 3×10^{20} eV gamma ray shower calculated with Landau-Pomeranchuk-Migdal (dashed) and Bethe-Heitler (solid) electromagnetic cross sections. The central line shows the average shower profile and the upper and lower lines show 1 σ deviations — not visible for the BH case, where lines overlap. The experimental shower profile is shown with the data points. It does not fit the profile of a photon shower.

observed shower profile does fit that of a primary proton, or, possibly, that of a nucleus. The shower profile information is sufficient, however, to conclude that the event is unlikely to be of photon origin.

2. The same conclusion is reached for the Yakutsk event that is characterized by a huge number of secondary muons, inconsistent with an electromagnetic cascade initiated by a gamma ray.

3. The AGASA collaboration claims evidence for "point" sources above 10 EeV. The arrival directions are however smeared out in a way consistent with primaries deflected by the galactic magnetic field. Again, this indicates charged primaries and excludes photons.

4. Finally, a recent reanalysis of the Haverah Park disfavors photon origin of the primaries [4].

Neutrino primaries are definitely ruled out. Standard model neutrino physics is understood, even for EeV energy. The average x of the parton mediating the neutrino interaction is of order $x \sim \sqrt{M_W^2/s} \sim 10^{-6}$ so that the perturbative result for the neutrino-nucleus cross section is calculable from measured HERA structure functions. Even at 100 EeV a reliable value of the cross section can be obtained based on QCD-inspired extrapolations of the structure function. The neutrino cross section is known to better than an order of magnitude. It falls 5 orders of magnitude short of the strong cross sections required to make a neutrino interact in the upper atmosphere to create an air shower.

Could EeV neutrinos be strongly interacting because of new physics? In theories with TeV-scale gravity, one can imagine that graviton exchange dominates all interactions and thus erases the difference between quarks and neutrinos at the energies under consideration. The actual models performing this feat require a fast turn-on of the cross section with energy that violates S-wave unitarity [13, 14, 15, 16, 17, 18, 19, 20, 21, 22, 23].

We have exhausted the possibilities. Neutrons, muons and other candidate primaries one may think of are unstable. EeV neutrons barely live long enough to reach us from sources at the edge of our galaxy.

D. A Three Prong Assault on the Cosmic Ray Puzzle

We conclude that, where the highest energy cosmic rays are concerned, both the accelerator mechanism and the particle physics are enigmatic. The mystery has inspired a worldwide effort to tackle the problem with novel experimentation in three complementary areas of research: air shower detection, atmospheric Cerenkov astronomy and underground neutrino astronomy. While some of the future instruments have additional missions, all are likely to have a major impact on cosmic ray physics.

1. Giant Cosmic Ray Detectors

With super-GZK fluxes of the order of a single event per square kilometer, per century, the outstanding problem is the lack of statistics; see Fig. 2 and Fig. 3. In the next five years, a qualitative improvement can be expected from the operation of the HiRes fluorescence detector in Utah. With improved instrumentation yielding high quality data from 2 detectors operated in coincidence, the interplay between sky transparency and energy measurement can be studied in detail. We can safely anticipate that the existence of super-GZK cosmic rays will be conclusively demonstrated by using the instrument's calorimetric measurements. A mostly Japanese collaboration has proposed a next-generation fluorescence detector, the Telescope Array.

The Auger air shower array is confronting the low rate problem with a huge collection area covering 3000 square kilometers on an elevated plain in Western Argentina. The instrumentation consists of 1600 water Cerenkov detectors spaced by 1.5 km. For calibration, about 15 percent of the showers occurring at night will be viewed by 3 HiRes-style fluorescence detectors. The detector is expected to observe several thousand events per year above 10 EeV and tens above 100 EeV. Exact numbers will depend on the detailed shape of the observed spectrum which is, at present, a matter of speculation.

2. Gamma rays from Cosmic Accelerators

An alternative way to identify the source(s) of the highest energy cosmic rays is illustrated in Fig. 5. The cartoon draws our attention to the fact that cosmic accelerators are also cosmic beam dumps which produce secondary photon and neutrino beams. Accelerating particles to TeV energy and above requires relativistic, massive bulk flows. These are likely to originate from the exceptional gravitational forces associated with dense cores of exploding stars, inflows onto supermassive black holes at the centers of active galaxies, annihilating black holes or neutron stars. In such situations, accelerated particles are likely to pass through intense radiation fields or dense clouds of gas surrounding the black hole. This leads to the production of secondary photons and neutrinos that accompany the primary cosmic ray beam. An example of an electromagnetic beam dump is the UV radiation field that surrounds the central black hole of active galaxies. The target material, whether a gas of particles or of photons, is likely to be tenuous enough that the primary beam and the photon beam are only partially attenuated. However, shrouded sources from which only neutrinos can emerge, as in terrestrial beam dumps at CERN and Fermilab, are also a possibility.

The astronomy event of the 21st century could be the simultaneous observation of TeV-

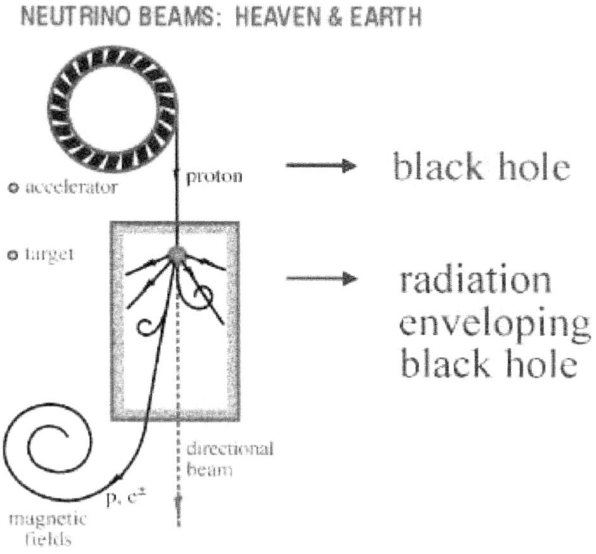

FIG. 5: Diagram of cosmic accelerator and beam dump. See text for discussion.

gamma rays, neutrinos and gravitational waves from cataclysmic events associated with the source of the highest energy cosmic rays.

We first concentrate on the possibility of detecting high-energy photon beams. After two decades, ground-based gamma ray astronomy has become a mature science [24, 25, 26, 27, 28, 29]. A large mirror, viewed by an array of photomultipliers, collects the Cerenkov light emitted by air showers and images the showers in order to determine the arrival direction and the nature of the primary particle. These experiments have opened a new window in astronomy by extending the photon spectrum to 20 TeV, and possibly beyond. Observations have revealed spectacular TeV-emission from galactic supernova remnants and nearby quasars, some of which emit most of their energy in very short bursts of TeV-photons.

But there is the dog that didn't bark. No evidence has emerged for the π^0 origin of TeV radiation. Therefore, no cosmic ray sources have yet been identified. Dedicated searches for photon beams from suspected cosmic ray sources, such as the supernova remnants IC433 and γ-Cygni, came up empty handed. While not relevant to the topic covered by this paper, supernova remnants are theorized to be the sources of the bulk of the cosmic rays that are of galactic origin. However, the evidence is still circumstantial.

The field of gamma ray astronomy is buzzing with activity to construct second-generation instruments. Space-based detectors are extending their reach from GeV to TeV energy with AMS and, especially, GLAST, while the ground-based Cerenkov collaborations are designing instruments with lower thresholds. Soon, both techniques should generate overlapping measurements in the $10-10^2$ GeV energy range. All ground-based air Cerenkov experiments aim at lower threshold, better angular and energy resolution, and a longer duty cycle. One can, however, identify three pathways to reach these goals:

1. larger mirror area, exploiting the parasitic use of solar collectors during nighttime (CELESTE, STACEY and SOLAR II) [30],

2. better, or rather, ultimate imaging with the 17m MAGIC mirror, [31]

3. larger field of view and better pointing and energy measurement using multiple telescopes (VERITAS, HEGRA and HESS).

The Whipple telescope pioneered the atmospheric Cerenkov technique. VERITAS [32] is an array of 9 upgraded Whipple telescopes, each with a field of view of 6 degrees. These can be operated in coincidence for improved angular resolution, or be pointed at 9 different 6 degree bins in the night sky, thus achieving a large field of view. The HEGRA collaboration [33] is already operating four telescopes in coincidence and is building an upgraded facility with excellent viewing and optimal location near the equator in Namibia.

There is a dark horse in this race: Milagro [34]. The Milagro idea is to lower the threshold of conventional air shower arrays to 100 GeV by instrumenting a pond of five million gallons of ultra-pure water with photomultipliers. For time-varying signals, such as bursts, the threshold may be even lower.

3. Neutrinos from Cosmic Accelerators

How many neutrinos are produced in association with the cosmic ray beam? The answer to this question, among many others [35, 36], provides the rational for building kilometer-scale neutrino detectors.

Let's first consider the question for the accelerator beam producing neutrino beams at an accelerator laboratory. Here the target absorbs all parent protons as well as the muons, electrons and gamma rays (from $\pi^0 \to \gamma + \gamma$) produced. A pure neutrino beam exits the dump. If nature constructed such a "hidden source" in the heavens, conventional astronomy will not reveal it. It cannot be the source of the cosmic rays, however, for which the dump must be partially transparent to protons.

In the other extreme, the accelerated proton interacts, thus producing the observed high-energy gamma rays, and subsequently escapes the dump. We refer to this as a transparent source. Particle physics directly relates the number of neutrinos to the number of observed cosmic rays and gamma rays[37]. Every observed cosmic ray interacts once, and only once, to produce a neutrino beam determined only by particle physics. The neutrino flux for such a transparent cosmic ray source is referred to as the Waxman-Bahcall flux [38, 39, 40, 41] and is shown as the horizontal lines labeled "W&B" in Fig. 6. The calculations is valid for $E \simeq 100 \, \text{PeV}$. If the flux is calculated at both lower and higher cosmic ray energies, however, larger values are found. This is shown as the non-flat line labeled "transparent" in Fig. 6. On the lower side, the neutrino flux is higher because it is normalized to a larger cosmic ray flux. On the higher side, there are more cosmic rays in the dump to produce neutrinos because the observed flux at Earth has been reduced by absorption on microwave photons, the GZK-effect. The increased values of the neutrino flux are also shown in Fig. 6. The gamma ray flux of π^0 origin associated with a transparent source is qualitatively at the level of observed flux of non-thermal TeV gamma rays from individual sources[37].

Nothing prevents us, however, from imagining heavenly beam dumps with target densities somewhere between those of hidden and transparent sources. When increasing the target photon density, the proton beam is absorbed in the dump and the number of neutrino-producing protons is enhanced relative to those escaping the source as cosmic rays. For the extreme source of this type, the observed cosmic rays are all decay products of neutrons with larger mean-free paths in the dump. The flux for such a source is shown as the upper

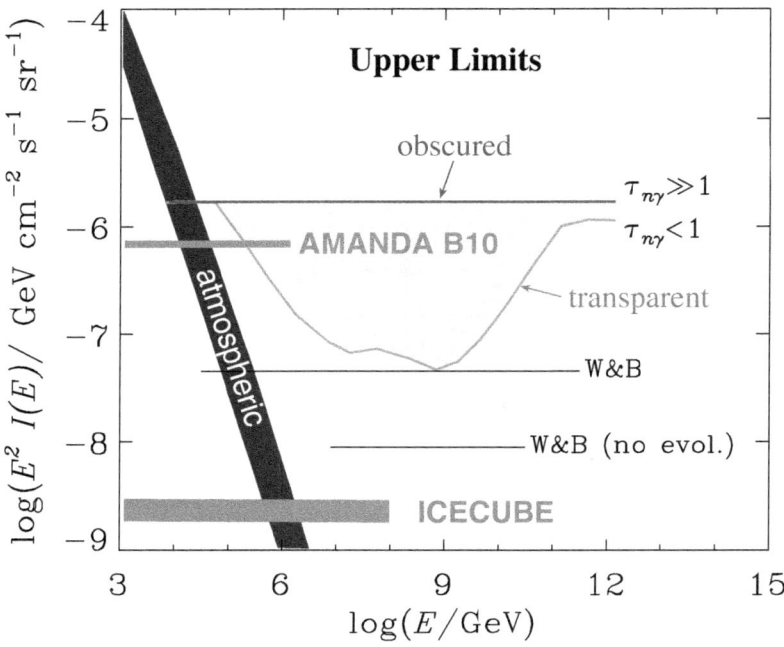

FIG. 6: The neutrino flux from compact astrophysical accelerators. Shown is the range of possible neutrino fluxes associated with the the highest energy cosmic rays. The lower line, labeled "transparent", represents a source where each cosmic ray interacts only once before escaping the object. The upper line, labeled "obscured", represents an ideal neutrino source where all cosmic rays escape in the form of neutrons. Also shown is the ability of AMANDA and IceCube to test these models.

horizontal line in Fig. 6.

The above limits are derived from the fact that theorized neutrino sources do not overproduce cosmic rays. Similarly, observed gamma ray fluxes constrain potential neutrino sources because for every parent charged pion ($\pi^\pm \to l^\pm + \nu$), a neutral pion and two gamma rays ($\pi^0 \to \gamma + \gamma$) are produced. The electromagnetic energy associated with the decay of neutral pions should not exceed observed astronomical fluxes. These calculations must take into account cascading of the electromagnetic flux in the background photon and magnetic fields. A simple argument relating high-energy photons and neutrinos produced by secondary pions can still be derived by relating their total energy and allowing for a steeper photon flux as a result of cascading. Identifying the photon fluxes with those of non-thermal TeV photons emitted by supernova remnants and blazers, we predict neutrino fluxes at the same level as the Waxman-Bahcall flux. It is important to realize however that there is no evidence that these are the decay products of π^0's. The sources of the cosmic rays have not been revealed by photon or proton astronomy [42, 43, 44, 45].

For neutrino detectors to succeed they must be sensitive to the range of fluxes covered in Fig. 6. The AMANDA detector has already entered the region of sensitivity and is eliminating specific models which predict the largest neutrino fluxes within the range of values allowed by general arguments. The IceCube detector, now under construction, is sensitive to the full range of beam dump models, whether generic as or modeled as active galaxies or gamma ray bursts. IceCube will reveal the sources of the cosmic rays or derive an upper limit that will qualitatively raise the bar for solving the cosmic ray puzzle. The situation could be nothing but desperate with the escape to top-down models being cut off by the accumulating evidence that the highest energy cosmic rays are not photons. In top-down models, decay products predominantly materialize as quarks and gluons that materialize as jets of neutrinos and photons and very few protons. We will return to top-down models at the end of this review.

II. HIGH-ENERGY NEUTRINO TELESCOPES

A. Observing High-energy Neutrinos

Although details vary from experiment to experiment, high-energy neutrino telescopes consist of strings of photo-multiplier tubes (PMT) distributed throughout a natural Cerenkov medium such as water or ice. Typical spacing of PMT is 10-20 meters along a string with string spacing of 30-100 meters. Such experiments can observe neutrinos of different flavors over a wide range of energies by using a variety of methods:

- Muon neutrinos that interact via charged current interactions produce a muon (along with a visible hadronic shower if the neutrino is of sufficient energy). The muon travels through the medium producing Cerenkov radiation which is detected by an array of PMT. The timing, amplitude (number of Cerenkov photons) and topology of the PMT signals is used to reconstruct the muon's path. The muon energy threshold for such a reconstruction is typically in the range of 10-100 GeV.

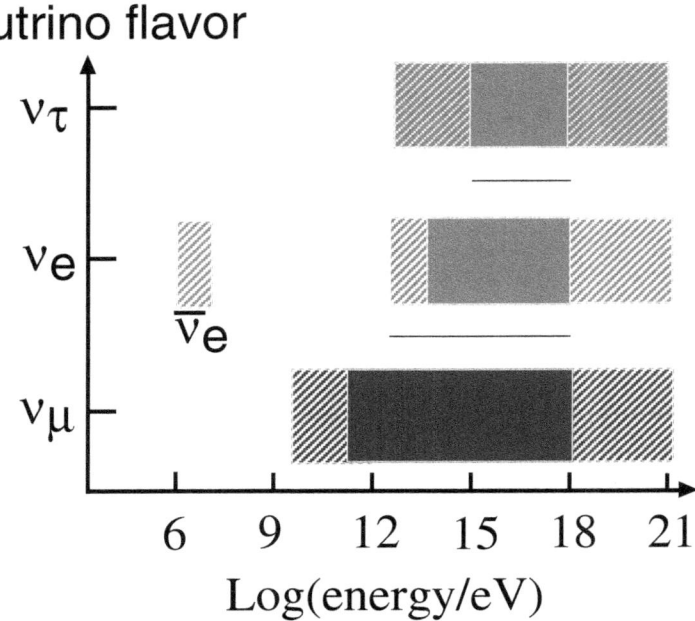

FIG. 7: Although IceCube detects neutrinos of any flavor, at TeV-EeV energies, it can identify their flavor and measure their energy in the ranges shown. Filled areas: particle identification, energy, and angle. Shaded areas: energy and angle.

To be detected, a neutrino must interact via charged current and produce a muon with sufficient range to reach the detector. The probability of detection is therefore the product of the interaction probability (or the inverse interaction length $\lambda_\nu^{-1} = n\sigma_\nu$) and the range of the muon R_μ:

$$P_{\nu \to \mu} \simeq n\sigma_\nu R_\mu \qquad (11)$$

where n is the number density of target nucleons, σ_ν is the charged current interaction cross section [48] and the range is $R_\mu \simeq 5\,\mathrm{m}$ per GeV for low energy muons. The muon range is determined by catastrophic energy loss (brehmsstrahlung, pair production and deep inelastic scattering) for muons with energies exceeding $\sim 500\,\mathrm{GeV}$ [46, 47].

- Muon, tau or electron neutrinos which interact via charged or neutral current interactions produce showers which can be observed when the interaction occurs within or close to the detector volume. Even the highest energy showers penetrate water or ice less than 10 m, a distance short compared to the typical spacing of the PMT. The Cerenkov light emitted by shower particles, therefore, represents a point source of light as viewed by the array. The radius over which PMT signals are produced is 250 m for a 1 PeV shower; this radius grows or decreases by approximately 50 m with every decade of shower energy. The threshold for showers is generally higher than for muons which limits neutral current identification for lower energy neutrinos. The probability for a neutrino to interact within the detector's effective area and to generate a shower within its volume is approximately given by:

$$P_{\nu \to \mu} \simeq n \sigma_\nu L \tag{12}$$

where σ_ν is the charged+neutral current interaction cross section, L is the length of the detector along the path of the neutrino and n, again, is the number density of target nucleons.

- Tau neutrinos are more difficult to detect but produce spectacular signatures at PeV energies. The identification of charged current tau neutrino events is made by observing one of two signatures: double bang events [49, 50, 51] and lollypop events [52, 53]. Double bang events occur when a tau lepton is produced along with a hadronic shower in a charged current interaction within the detector volume and the tau decays producing a electromagnetic or hadronic shower before exiting the detector (as shown in Fig. 8). Below a few PeV, the two showers cannot be distinguished. Lollypop events occur when only the second of the two showers of a double bang event occurs within the detector volume and a tau lepton track is identified entering the shower over several hundred meters. The incoming τ can be clearly distinguished from a muon. A muon initiating a PeV shower would undergo observable catastrophic energylosses. Lollypop events are useful only at several PeV energies are above. Below this energy, tau tracks are not long enough to be identified.

  A feature unique to tau neutrinos is that they are not depleted in number by absorption in the earth. Tau neutrinos which interact producing a tau lepton generate another tau neutrino when the tau lepton decays, thus only degrading the energy of the neutrino [54, 55, 56, 57].

- Although MeV scale neutrinos are far below the energies required to identify individual events, large fluxes of MeV electron anti-neutrinos interacting via charged current could be detected by observing higher counting rates of individual PMT over a time window of several seconds. The enhancement rate in a single PMT will be buried in dark noise of that PMT. However, summing the signals from all PMT over a short time window can reveal significant excesses, for instance form a galactic supernova.

With these signatures, neutrino astronomy can study neutrinos from the MeV range to the highest known energies ($\sim 10^{20}$eV).

B. Large Natural Cerenkov Detectors

A new window in astronomy is upon us as high-energy neutrino telescopes see first light [58]. Although neutrino telescopes have multiple interdisciplinary science missions, the search for the sources of the highest-energy cosmic rays stands out because it most directly identifies the size of the detector required to do the science [46, 47]. For guidance in estimating expected signals, one makes use of data covering the highest-energy cosmic rays in Fig. 2 and Fig. 3 as well as known sources of non-thermal, high-energy gamma rays. Estimates based on this information suggest that a kilometer-scale detector is needed to see neutrino signals as previously discussed.

The same conclusion is reached using specific models. Assume, for instance, that gamma ray bursts (GRB) are the cosmic accelerators of the highest-energy cosmic rays. One can calculate from textbook particle physics how many neutrinos are produced when the particle beam coexists with the observed MeV energy photons in the original fireball. We thus predict the observation of 10–100 neutrinos of PeV energy per year in a detector with a square kilometer effective area. GRB are an example of a generic beam dump associated with the highest energy cosmic rays. We will work through this example in some detail in later sections. In general, the potential scientific payoff of doing neutrino astronomy arises from the great penetrating power of neutrinos, which allows them to emerge from dense inner regions of energetic sources.

The strong scientific motivations for a large area, high-energy neutrino observatory lead to the formidable challenges of developing effective, reliable and affordable detector technology. Suggestions to use a large volume of deep ocean water for high-energy neutrino astronomy were made as early as the 1960s. Today, with the first observation of neutrinos in the Lake

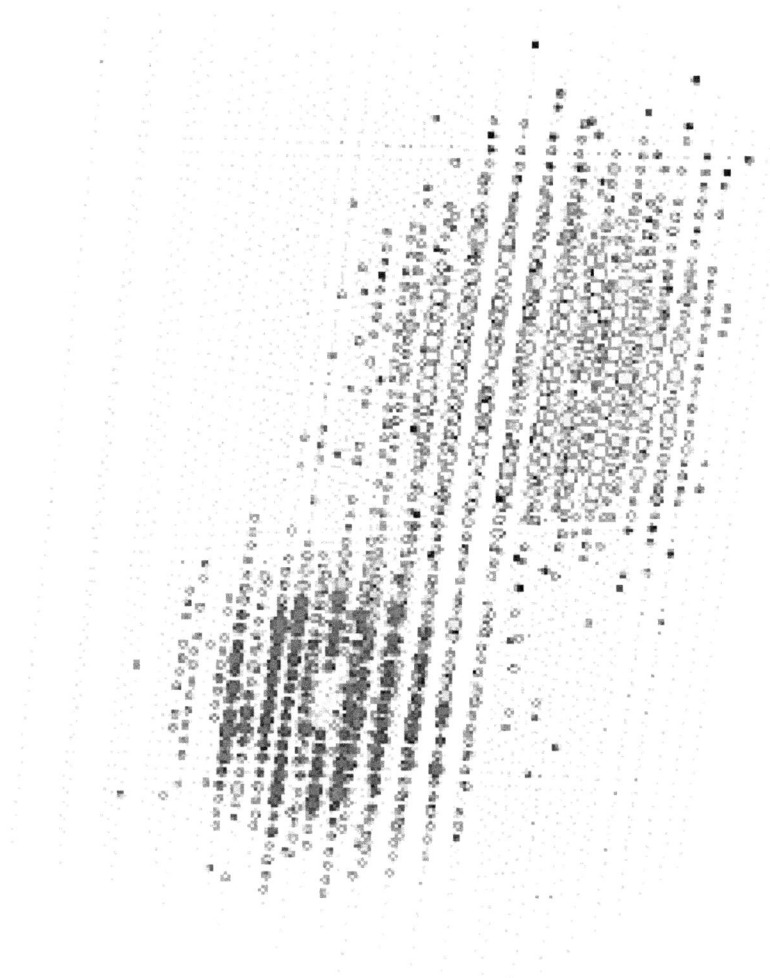

FIG. 8: Simulation of an ultra high-energy tau lepton generated by the interaction of a 10 PeV tau neutrino (first shower), followed by the decay of the secondary tau lepton (second shower). The shading represents the time sequence of the hits. The size of the dots corresponds to the number of photons detected by the individual photomultipliers.

Baikal and the South Pole neutrino telescopes, there is optimism that the technological challenges of building neutrino telescopes have been met.

Launched by the bold decision of the DUMAND collaboration to construct such an in-

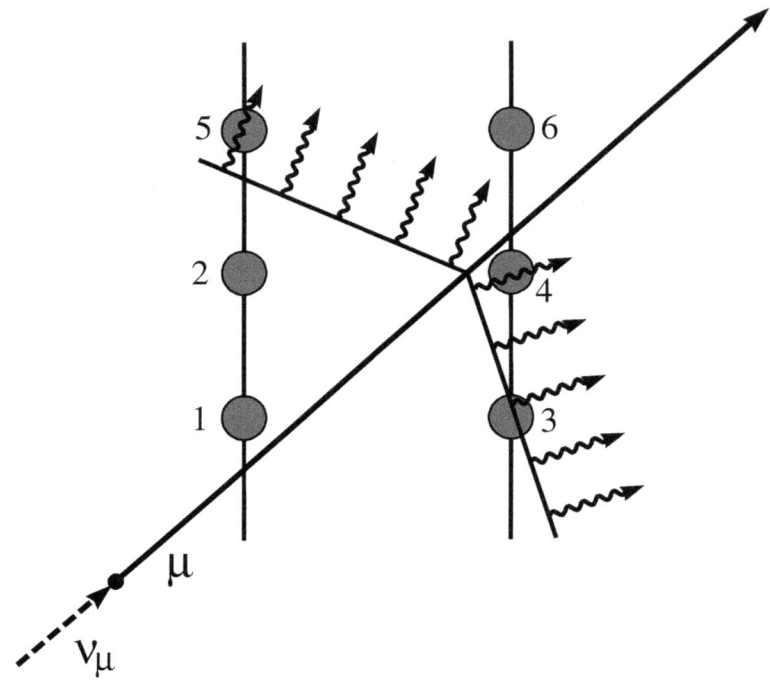

FIG. 9: The arrival times of the Cerenkov photons in 6 optical sensors determine the direction of the muon track.

strument, the first generation of neutrino telescopes is designed to reach a large telescope area and detection volume for a neutrino threshold of order 10 GeV [59, 60, 61]. This relatively low threshold permits calibration of the novel instrumentation on the known flux of atmospheric neutrinos. The architecture is optimized for reconstructing the Cerenkov light front radiated by an up-going, neutrino-induced muon. Up-going muons must be identified in a background of down-going, cosmic ray muons which are more than 10^5 times more frequent for a depth of ∼1–2 kilometers. The earth is used as a filter to screen out the background of down-going cosmic ray muons. This makes neutrino detection possible over the hemisphere of sky faced by the bottom of the detector.

The optical requirements on the detector medium are severe. A large absorption length is needed because it determines the required spacing of the optical sensors and, to a significant extent, the cost of the detector. A long scattering length is needed to preserve the geometry of the Cerenkov pattern. Nature has been kind and offered ice and water as natural Cerenkov media. Their optical properties are, in fact, complementary. Water and ice have similar attenuation length, with the roles of scattering and absorption reversed. Optics seems, at present, to drive the evolution of ice and water detectors in predictable directions: towards very large telescope area in ice exploiting the long absorption length, and towards lower threshold and good muon track reconstruction in water exploiting the long scattering length.

1. Baikal, ANTARES, Nestor and NEMO: Northern Water

Whereas the science is compelling, we now turn to the challenge of developing effective detector technology. With the termination of the pioneering DUMAND experiment, the efforts in water are, at present, spearheaded by the Baikal experiment [62, 63, 64, 65]. The Baikal Neutrino Telescope is deployed in Lake Baikal, Siberia, 3.6 km from shore at a depth of 1.1 km. An umbrella-like frame holds 8 strings, each instrumented with 24 pairs of 37-cm diameter *QUASAR* photomultiplier tubes. Two PMT are required to trigger in coincidence in order to suppress the large background rates produced by natural radioactivity and bioluminescence in individual PMT. Operating with 144 optical modules (OM) since April 1997, the *NT-200* detector was completed in April 1998 with 192 OM. Due to unstable electronics only ∼ 60 channels took data during 1998. Nevertheless 35 neutrino-induced upgoing muons were identified in the first 234 live days of data; see Fig. 10 for a 70 day sample. The neutrino events are isolated from the cosmic ray muon background by imposing a restriction on the chi-square of the fit of measured photon arrival times and amplitudes to a Cherenkov cone, and by requiring consistency between the reconstructed trajectory and the spatial locations of the OMs reporting signals. In order to guarantee a minimum lever arm for track fitting, they only consider events with a projection of the most distant channels on the track larger than 35 meters. This does, of course, result in a higher energy threshold. Agreement with the expected atmospheric neutrino flux of 31 events shows that the Baikal detector is understood. Stability and performance of the detector have improved in 1999 and 2000 data taking [65].

The Baikal site is competitive with deep oceans, although the smaller absorption length of Cerenkov light in lake water requires a somewhat denser spacing of the OMs. This does,

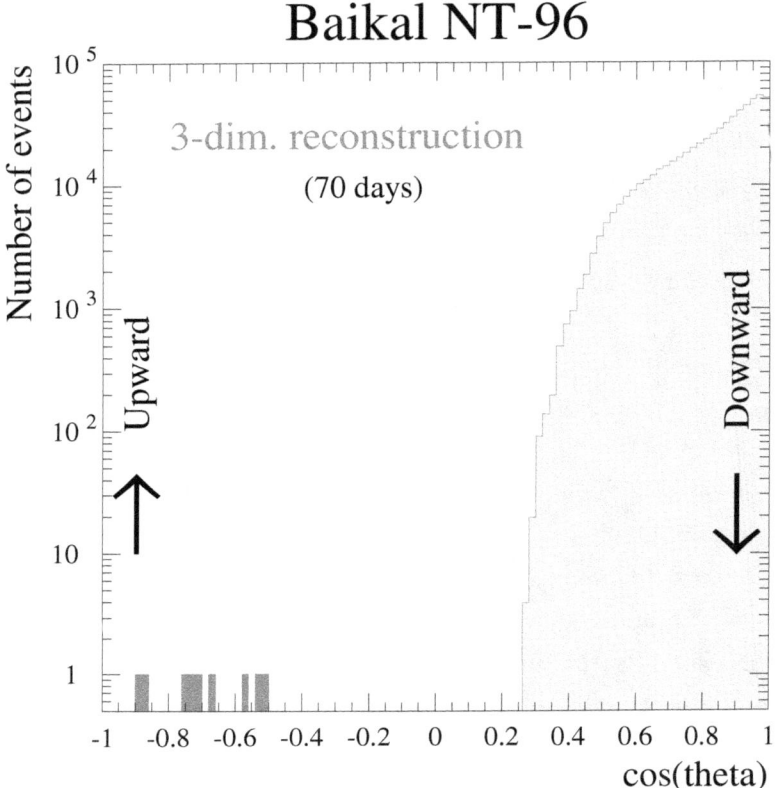

FIG. 10: Angular distribution of muon tracks in the Lake Baikal experiment after the cuts described in the text.

however, result in a lower threshold which is a definite advantage, for instance for oscillation measurements and WIMP searches. They have shown that their shallow depth of 1 kilometer does not represent a serious drawback. A significant advantage is that the site has a seasonal ice cover which allows reliable and inexpensive deployment and repair of detector elements.

In the following years, *NT-200* will be operated as a neutrino telescope with an effective area between 10^3 and $5 \times 10^3 \, \text{m}^2$, depending on energy. Presumably too small to detect neutrinos from extraterrestrial sources, *NT-200* will serve as the prototype for a larger telescope. For instance, with 2000 OMs, a threshold of 10 to 20 GeV and an effective area of 5×10^4 to $10^5 \, \text{m}^2$, an expanded Baikal telescope could fill the gap between present underground detectors and planned high threshold detectors of cubic kilometer size. Its key advantage would be low energy threshold.

The Baikal experiment represents a proof of concept for future deep ocean projects that have the advantage of larger depth and optically superior water. Their challenge is to find reliable and affordable solutions to a variety of technological challenges for deploying a deep underwater detector. Several groups are confronting the problem; both NESTOR and ANTARES are developing rather different detector concepts in the Mediterranean.

The NESTOR collaboration [66, 67, 68], as part of a series of ongoing technology tests, is testing the umbrella structure which will hold the OMs. They have already deployed two aluminum "floors", 34 m in diameter, to a depth of 2600 m. Mechanical robustness was demonstrated by towing the structure, submerged below 2000 m, from shore to the site and back. These tests should soon be repeated with two fully instrumented floors. The cable connecting the instrument to the counting house on shore has been deployed. The final detector will consist of a tower of 12 six-legged floors vertically separated by 30 m. Each floor contains 14 OMs with four times the photocathode area of the commercial 8 inch photomultipliers used by AMANDA and ANTARES.

The detector concept is patterned along the Baikal design. The symmetric up/down orientation of the OMs will result in uniform angular acceptance and the relatively close spacings will result in a low energy threshold. NESTOR does have the advantage of a superb site off the coast of Southern Greece, possibly the best in the Mediterranean. The detector can be deployed below 3.5 km relatively close to shore. With the attenuation length peaking at 55 m near 470 nm, the site is optically similar to that of the best deep water sites investigated for neutrino astronomy.

The ANTARES collaboration [69, 70, 71] is currently constructing a neutrino telescope at a 2400 m deep Mediterranean site off Toulon, France. The site is a trade-off between acceptable optical properties of the water and easy access to ocean technology. Their detector concept requires remotely operated vehicles for making underwater connections. Results on water quality are very encouraging with an absorption length of 40 m at 467 nm and 20 m at 375 nm, and a scattering length exceeding 100 m at both wavelengths. Random noise, exceeding 50 khz per OM, is eliminated by requiring coincidences between neighboring OMs, as is done in the Lake Baikal design. Unlike other water experiments, they will point all photomultipliers sideways or down in order to avoid the effects of biofouling. The problem is significant at the Toulon site, but only affects the upper pole region of the OM. Relatively weak intensity and long duration bioluminescence results in an acceptable deadtime of the detector. They have demonstrated their capability to deploy and retrieve a string, and have reconstructed down-going muons with 8 OMs deployed on the test string.

The ANTARES detector will consist of 13 strings, each equipped with 30 stories and 3 PMT per story. This detector will have an area of about $3 \times 10^4 \, \text{m}^2$ for 1 TeV muons — similar to AMANDA-II — and is planned to be fully deployed by the end of 2004. The electro-optical cable linking the underwater site to the shore was successfully deployed in October 2001.

NEMO, a new R&D initiative based in Catania, Sicily has been mapping Mediterranean sites, studying mechanical structures and low power electronics. One hopes that with a successful pioneering neutrino detector of $10^{-3} \, \text{km}^3$ in Lake Baikal and a forthcoming $10^{-2} \, \text{km}^3$ detector near Toulon, the Mediterranean effort will converge on a $10^{-1} \, \text{km}^3$ detector, possibly at the NESTOR site [72, 73]. For neutrino astronomy to become a viable science, several projects will have to succeed in addition to AMANDA. Astronomy, whether in the optical or in any other wave-band, thrives on a diversity of complementary instruments, not on "a single best instrument".

2. AMANDA: Southern Ice

Construction of the first-generation AMANDA-B10 detector [74, 75, 76, 77, 78] was completed in the austral summer 96–97. It consists of 302 optical modules deployed at a depth of 1500–2000 m; see Fig. 11. Here the optical modules consist of 8-inch photomultiplier tubes and are controlled by passive electronics. Each is connected to the surface by a cable that transmits the high voltage as well as the anode current of a triggered photomultiplier. The instrumented volume and the effective telescope area of this instrument matches those of the ultimate DUMAND Octagon detector which, unfortunately, could not be completed.

Depending on depth, the absorption length of blue and UV light in the ice varies between 85 and 225 meters. The effective scattering length, which combines the mean-free path λ with the average scattering angle θ as $\frac{\lambda}{(1-\langle \cos\theta \rangle)}$, varies from 15 to 40 meters [79]. Because the absorption length of light in the ice is very long and the scattering length relatively short, many photons are delayed by scattering. In order to reconstruct the muon track, maximum likelihood methods are used, which take into account the scattering and absorption

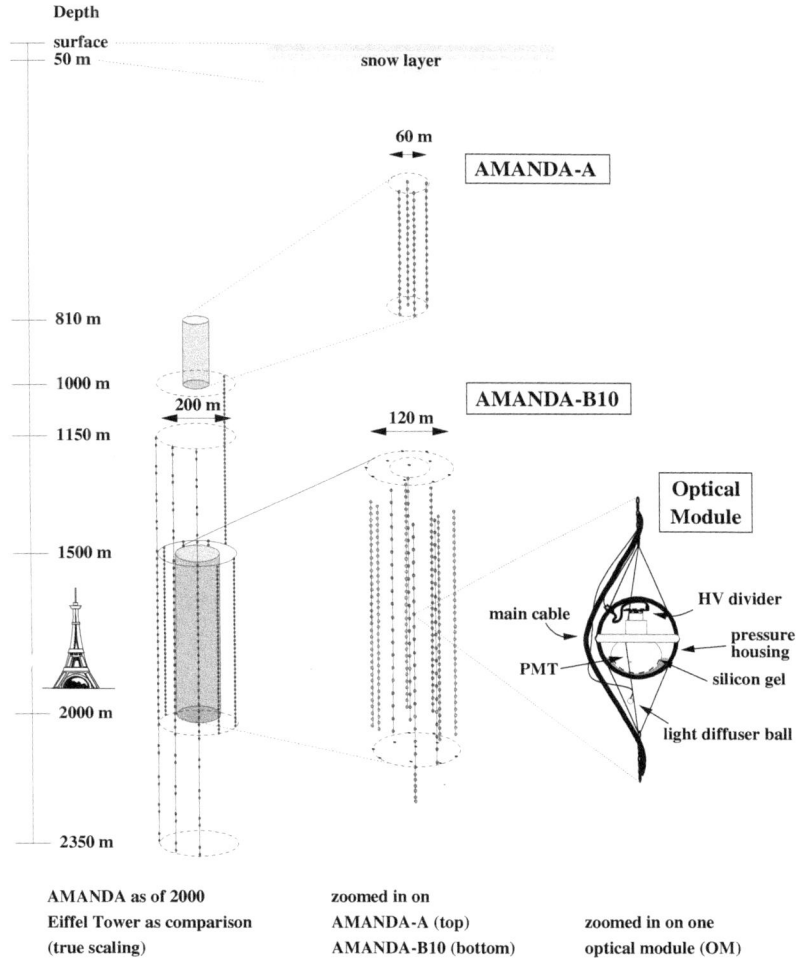

FIG. 11: The AMANDA detector and a schematic diagram of an optical module. Each dot represents an optical module. The modules are separated by 20 meters in the inner strings 1-4, and by 10 meters in the outer strings 5-10.

of photons as determined from calibration measurements [74]. A Bayesian formulation of the likelihood [80], which accounts for the much larger rate of down-going cosmic-ray muon tracks relative to up-going signal, has been particularly effective in decreasing the chance for a down-going muon to be misreconstructed as up-going.

Other types of events that might appear to be up-going muons must also be considered and eliminated. Rare cases, such as muons which undergo catastrophic energy loss, for instance through bremsstrahlung, or that are coincident with other muons, must be investigated. To this end, a series of requirements or quality criteria, based on the characteristic time and spatial pattern of photons associated with a muon track and the response of the detector, are applied to all events that, in the first analysis, appear to be up-going muons. For example, an event which has a large number of optical modules hit by photons unscattered (relative to the expected Cerenkov times of the reconstructed track) has a high quality. By making these requirements (or "cuts") increasingly selective, they eliminate more of the background of false up-going events while still retaining a significant fraction of the true up-going muons, i.e., the neutrino signal. Two different and independent analyses of the same data covering 138 days of observation in 1997 have been undertaken. These analyses yielded comparable numbers of up-going muons (153 in analysis A, 188 in analysis B). Comparison of these results with their respective Monte Carlo simulations shows that they are consistent with each other in terms of the numbers of events, the number of events in common, and, as discussed below, the expected properties of atmospheric neutrinos.

In Fig. 12, from analysis A, the experimental events are compared to simulations of background and signal as a function of the (identical) quality requirements placed on the three types of events: experimental data, simulated up-going muons from atmospheric neutrinos, and a simulated background of down-going cosmic ray muons. For simplicity in presentation, the levels of the individual types of cuts have been combined into a single parameter representing the overall event quality, and the comparison is made in the form of ratios. Fig. 12 shows events for which the quality level is 4 and higher. As the quality level is increased further, the ratios of simulated background to experimental data and experimental data to simulated signal both continue their rapid decrease, the former toward zero and the latter toward unity. Over the same range, the ratio of experimental data to the simulated sum of background and signal remains near unity. At an event quality of 6.9 there are 153 events in the sample of experimental data and the ratio to predicted signal is 0.7. The conclusions are that (1) the quality requirements have reduced the events from misreconstructed down-going muons in the experimental data to a negligible fraction of the signal and that (2) the experimental data behave in the same way as the simulated atmospheric neutrino signal for events that pass the stringent cuts. They estimate that the remaining signal is contaminated by instrumental background at 15 ± 7 percent.

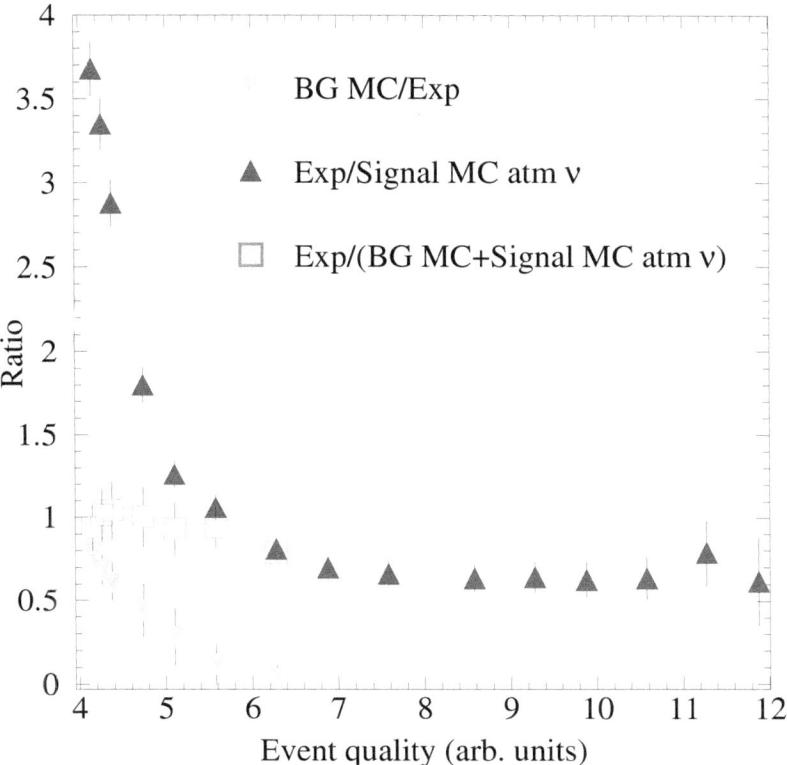

FIG. 12: Reconstructed muon events in AMANDA-B10 are compared to simulations of background cosmic ray muons (BG MC) and simulations of atmospheric neutrinos (Signal MC atm ν) as a function of "event quality", a variable indicating the severity of the cuts designed to enhance the signal. Note that the comparison is made in the form of ratios.

The estimated uncertainty on the number of events predicted by the signal Monte Carlo simulation (which includes uncertainties in the high-energy atmospheric neutrino flux, the sensitivity of the optical modules, and the precise optical properties of the ice) is $+40\%$ to -50%. The observed ratio of experiment to simulation (0.7) and the expectation (1.0) therefore agree within errors.

The shape of the zenith angle distribution from analysis B is compared to a simulation of the atmospheric neutrino signal in Fig. 13 in which the two distributions have been normalized to each other. The variation of the measured rate with zenith angle is reproduced by simulation to within the statistical uncertainty. Note that the tall geometry of the

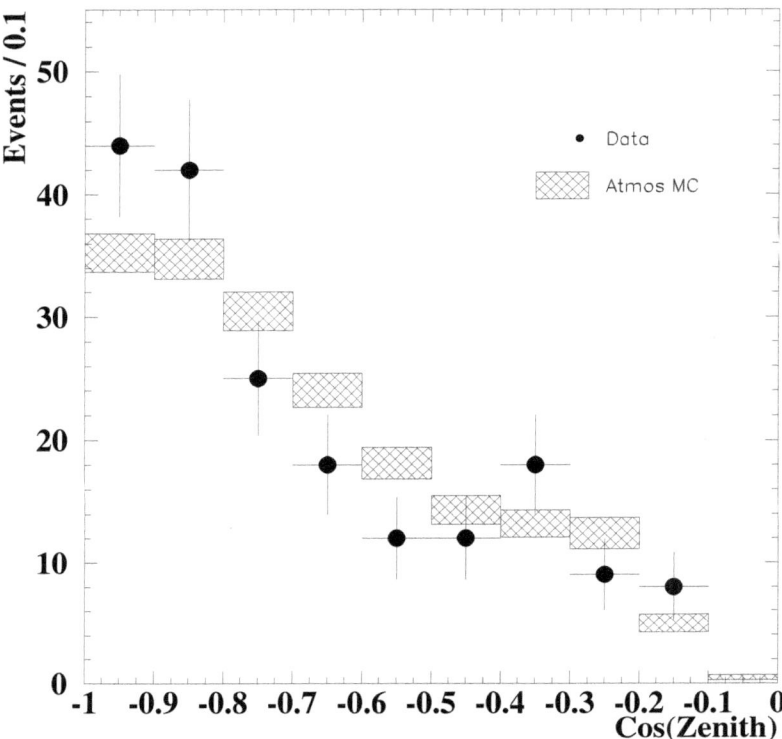

FIG. 13: Reconstructed zenith angle distribution for AMANDA-B10. The points mark the data and the shaded boxes a simulation of atmospheric neutrino events. The widths of the boxes indicate the error bars. The overall normalization of the simulation has been adjusted to match the data.

detector strongly influences the dependence on zenith angle in favor of more vertical muons.

Estimates of the energies of the up-going muons (based on simulations of the number of optical modules that participate in an event) indicate that the energies of these muons are in the range from 100 GeV to ~ 1 TeV. This is consistent with their atmospheric neutrino origin.

The agreement between simulation and experiment shown in Fig. 12 and 13, taken together with other comparisons of measured and simulated events, leads us to conclude that the up-going muon events observed by AMANDA are produced mainly by atmospheric neutrinos.

The arrival directions of the neutrinos observed in both analyses are shown in Fig. 14.

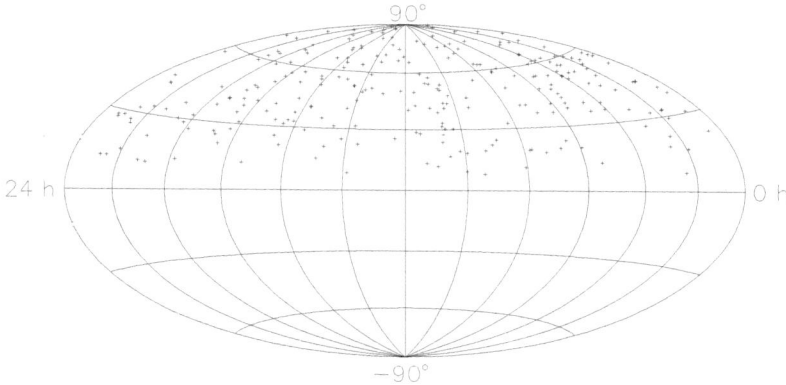

FIG. 14: Distribution in declination and right ascention of the up-going AMANDA-B10 events on the sky.

A statistical analysis indicates no evidence for point sources in this sample. An estimate of the energies of the up-going muons indicates that all events have energies consistent with an atmospheric neutrino origin. This corresponds to a level of sensitivity to a diffuse flux of high-energy extra-terrestrial neutrinos of order $dN/dE_\nu = 10^{-6} E_\nu^{-2}\,\mathrm{cm}^{-2}\,\mathrm{s}^{-1}\,\mathrm{sr}^{-1}\,\mathrm{GeV}^{-1}$, assuming an E^{-2} spectrum [81]. This upper limit excludes a variety of theoretical models which assume the hadronic origin of TeV photons from active galaxies and blazars. Searches for neutrinos from gamma ray bursts, magnetic monopoles, and for a cold dark matter signal from the center of the Earth yield limits comparable to or better than those from smaller underground neutrino detectors that have operated for a much longer period.

Data are being taken now with the larger array, AMANDA-II consisting of an additional 480 OMs.

3. IceCube: A Kilometer-Scale Neutrino Observatory

The IceCube project [82, 83] at the South Pole is a logical extension of the research and development work performed over the past several years by the AMANDA Collaboration. The optimized design for IceCube is an array of 4800 photomultiplier tubes each enclosed in a transparent pressure sphere to comprise an optical module similar to those in AMANDA. In the IceCube design, 80 strings are regularly spaced by 125 m over an area of approximately one square kilometer, with OMs at depths from 1.4 to 2.4 km below the surface. Each string consists of OMs connected electrically and mechanically to a long cable which brings OM signals to the surface. The array is deployed one string at a time. For each string, a enhanced hot-water drill melts a hole in the ice to a depth of about 2.4 km in less than 2 days. The drill is then removed from the hole and a string with 60 OMs vertically spaced by 17 m is deployed before the water re-freezes. The signal cables from all the strings are brought to a central location which houses the data acquisition electronics, other electronics, and computing equipment.

Each OM contains a 10 inch PMT that detects individual photons of Cerenkov light generated in the optically clear ice by muons and electrons moving with velocities near the speed of light.

Background events are mainly down-going muons from cosmic ray interactions in the atmosphere above the detector. The background is monitored for calibration purposes and background rejection by the IceTop air shower array covering the detector.

Signals from the optical modules are digitized and transmitted to the surface such that a photon's time of arrival at an OM can be determined to within less than 5 nanoseconds. The electronics at the surface determines when an event has occurred (e.g., that a muon traversed or passed near the array) and records the information for subsequent event reconstruction and analysis.

At the South Pole site (see Fig. 15), a computer system accepts the data from the event trigger via the data acquisition system. The event rate, which is dominated by down-going cosmic ray muons, is estimated to be 1–2 kHz. The technology that will be employed in IceCube has been developed, tested, and demonstrated in AMANDA deployments, in laboratory testing, and in simulations validated by AMANDA data. This includes the instrument architecture, technology, deployment, calibration, and scientific utilization of the proposed detector. There have been yearly improvements in the AMANDA system, especially in the OMs, and in the overall quality of the information obtained from the detector. In the 1999/2000 season, a string was deployed with optical modules containing readout electronics inside the OM. The information is sent digitally to the surface over twisted-pair electrical cable. This option eliminates the need for optical fiber cables and simplifies calibration of the detector elements. This digital technology is the baseline technology of IceCube. For more details, see Ref. [84].

The construction of neutrino telescopes is overwhelmingly motivated by their discovery

FIG. 15: The South Pole site, showing the residential dome and associated buildings, the skiway where planes land, the dark sector with the Martin A. Pomerantz Observatory in which the AMANDA electronics are housed, and a rough outline of where IceCube strings are to be placed.

potential in astronomy, astrophysics, cosmology and particle physics. To maximize this potential, one must design an instrument with the largest possible effective telescope area to overcome the neutrino's small cross section with matter, and the best possible angular and energy resolution to address the wide diversity of possible signals.

At this point in time, several of the new instruments (such as the partially deployed Auger array, HiRes, Magic, Milagro and AMANDA II) are less than one year from delivering results. With rapidly growing observational capabilities, one can realistically hope, almost 100 years after their discovery, the puzzling origin of the cosmic rays will be deciphered. The solution will almost certainly reveal unexpected astrophysics or particle physics.

C. EeV Neutrino Astronomy

At extremely high energies, new techniques can be used to detect astrophysical neutrinos. These include the detection of acoustic and radio signals induced by super-EeV neutrinos interacting in water, ice or salt domes, or the detection of horizontal air showers by large conventional cosmic ray experiments such as the Auger array.

Horizontal air showers are likely to be initiated by a neutrino because showers induced by primary cosmic rays are unlikely to penetrate the $\sim 36,000\,\text{g/cm}^2$ of atmosphere along the horizon. Isolated penetrating muons may survive but they can be experimentally separated from a shower initiated by a neutrino close to the detector. Horizontal air shower experiments can also use nearby mountains as a target, e.g. to observe the decay of tau leptons produced in charged current interactions in the moutain. The sensitivity of an air shower array to detect an ultra high-energy neutrino is described by its acceptance, expressed in units of km^3 water equivalent steradians ($\text{km}^3\text{we sr}$). Typically only showers with zenith angle greater than ~ 70 degrees can be identified as neutrinos. This corresponds to a slant depth of $\sim 2000\,\text{g/cm}^2$.

The acceptance of present air shower experiments, such as AGASA, is $\sim 1\,\text{km}^3\text{we sr}$ above $10^{10}\,\text{GeV}$, and significantly less at lower energies. Auger will achieve ten times greater acceptance at $10^9\,\text{GeV}$ and 50 times greater near $10^{12}\,\text{GeV}$. Nitrogen fluorescence experiments also have the capability to detect neutrinos as nearly horizontal air showers with space-based experiments such as EUSO and OWL extending the reach of Auger. At this point we should point out however that the actual event rates of these experiments are similar to those for IceCube. Although IceCubes energy resolution saturates at EeV energies, the neutrinos are still detected with rates competitive with the most ambitious horizontal air shower experiments; for a more detailed comparison see Ref.[85, 86].

Radio Cerenkov experiments detect the Giga-Hertz pulse radiated by shower electrons produced in the interaction of neutrinos in ice. Also, the moon, viewed by ground-based radio telescopes, has been used as a target [87]. Above a threshold of $\simeq 1\,\text{PeV}$, the large number of low energy($\simeq MeV$) photons in a shower will produce an excess of electrons over positrons by removing electrons from atoms by Compton scattering. These are the sources of coherent radiation at radio frequencies, i.e. above $\sim 100\,\text{MHz}$. The mechanism is now well understood. The characteristics and the power of the pulses have been measured by dumping a photon beam in sand [88]. The results agree with calculations [89].

While many proposals exist, the most extensive effort to develop a radio neutrino detector is RICE (Radio Ice Cerenkov Experiment), which is located in the shallow ice above the AMANDA detector [90]. It consists of an 18-channel array of radio receivers distributed within a $8 \times 10^6 \, \text{m}^3$ volume. The receivers, buried in the ice at depths of 100-300 meters, are sensitive over the range of 0.2-1 GHz, roughly corresponding to electron neutrinos with energy of several PeV and above. The ANITA collaboration proposes to fly a balloon-borne array of radio antennas on a circular flight over Antarctica. ANITA will detect earth-skimming neutrinos [91] producing signals emerging from the ice along the horizon [92]. With higher threshold but also greater effective area than RICE (about 1 million km^2), ANITA should be sensitive to GZK neutrinos after a lucky 30 day flight (or 3 normal flights of 10 days).

EeV neutrino-induced showers can also be detected by acoustic emission resulting from local heating of a dense medium. Existing arrays of hydrophones, built in the earth's oceans for military application, could be used for the hydro-acoustic detection of neutrinos with extremely high energies; for a recent review see [93].

III. COSMIC NEUTRINO SOURCES

A. A List of Cosmic Neutrino Sources

We have previously discussed generic cosmic ray producing beam dumps and their associated neutrino fluxes. We now turn to specific sources of high-energy neutrinos. The list of proposed sources is long and includes, but is not limited to:

- Gamma Ray Bursts (GRB)

  GRB, outshining the entire universe for the duration of the burst, are perhaps the best motivated source for high-energy neutrinos [94, 95, 96]. Although we do not yet understand the internal mechanisms that generate GRB, the relativistic fireball model provides us with a successful phenomenology accommodating observations. It is very likely that GRB are generated in some type of cataclysmic process involving dying massive stars. GRB may prove to be an excellent source of neutrinos with energies from MeV to EeV and above. As we shall demonstrate further on, their fluxes can be calculated in a relatively model independent fashion.

- Other Sources Associated with Stellar Objects

  Other theorized neutrino sources associated with compact objects include supernova remnants exploding into the interstellar medium [46, 47, 98, 99], X-ray binaries [46, 100, 101, 102], microquasars [47, 103, 104] and even the sun [46, 47, 105, 106], any of which could provide observable fluxes of high-energy neutrinos.

- Active Galactic Nuclei (AGN): Blazars

    Blazars, the brightest objects in the universe and the sources of TeV-energy gamma rays, have been extensively studied as potential neutrino sources. Blazar flares with durations ranging from months to less than an hour, are believed to be produced by relativistic jets projected from an extremely massive accreting black hole. Blazars may be the sources of the highest energy cosmic rays and, in association, provide observable fluxes of neutrinos from TeV to EeV energies.

- Neutrinos Associated with the Propagation of Cosmic Rays

    Very high-energy cosmic rays generate neutrinos in interactions with the cosmic microwave background [107, 108]. This cosmogenic flux is among the most likely sources of high-energy neutrinos, and the most straightforward to predict. Furthermore, cosmic rays interact with the Earth's atmosphere [109, 110] and with the hydrogen concentrated in the galactic plane [46, 47, 111, 112, 113] producing high-energy neutrinos. It has also been proposed that cosmic neutrinos themselves may produce cosmic rays and neutrinos in interactions with relic neutrinos $\nu + \nu_b \to Z$. This is called the Z-burst mechanism [114, 115, 116, 117, 118].

- Dark Matter, Primordial Black Holes, Topological Defects and Top-Down Models

    The vast majority of matter in the universe is dark with its particle nature not yet revealed. The lightest supersymmetric particle, or other Weakly Interacting Massive Particles (WIMPs) propsed as particle candidates for cold dark matter, should become gravitationally trapped in the sun, earth or galactic center. There, they annihilate generating high-energy neutrinos observable in neutrino telescopes [119, 120, 121, 122, 123, 124, 125]. Another class of dark matter candidates are superheavy particles with GUT-scale masses that may generate the ultra high-energy cosmic rays by decay or annihilation, as well as solve the dark matter problem. These will also generate a substantial neutrino flux [126, 127, 128, 130]. Extremely high-energy neutrinos are also predicted in a wide variety of top-down scenarios invoked to produce cosmic rays, including decaying monopoles, vibrating cosmic strings [131, 132] and Hawking radiation from primordial black holes [133, 134, 135].

Any of these sources may or may not provide observable fluxes of neutrinos. History testifies to the fact that we have not been particularly successful at predicting the phenomena invariably revealed by new ways of viewing the heavens. We do, however, know that cosmic rays exist and that nature accelerates particles to super-EeV energy. In this review we concentrate on neutrino fluxes associated with the highest energy cosmic rays. Even here the anticipated flux depends on our speculation regarding the source. We will work through three much-researched examples: GRB, AGN and decays of particles or defects associated with the GUT-scale. The myriad of speculations have been recently reviewed by Learned and Mannheim [47]. We concentrate here on neutrino sources associated with the observed cosmic rays and gamma rays.

B. Gamma Ray Bursts: A Detailed Example of a Generic Beam Dump

1. GRB Characteristics

Although there is no such thing as a typical gamma ray burst, observations of GRB indicate the following common characteristics:

- GRB are extremely luminous events, often releasing energy of order one solar mass in gamma rays. Typically, $L_\gamma \sim 10^{51}$ to 10^{54} erg/s is released over durations of seconds or tens of seconds. GRB are the most luminous sources in the universe.

- GRB produce a broken power-law spectrum of gamma rays with $\phi_\gamma \propto E_\gamma^{-2}$ for $E_\gamma \gtrsim$ 0.1-1 MeV and $\phi_\gamma \propto E_\gamma^{-1}$ for $E_\gamma \lesssim$ 0.1-1 MeV [136, 137].

- GRB are cosmological events. Redshifts exceeding z=4 have been measured [138, 139].

- GRB are rare. During it's operation, BATSE observed on average 1 burst per day within its field of view ($\sim 1/3$ of the sky). Assuming that the rate of GRB does not significantly change with cosmological time, this corresponds to one burst per galaxy per million years. If GRB are beamed, they may be more common.

- GRB produce afterglows of less energetic photons which extend long after the initial burst [140, 141, 142].

- The durations of GRB follow a bimodal distribution with peaks near two seconds and 20 seconds, although some GRB have durations ranging from milliseconds to 1000 seconds [143]. Variations in the spectra occur on the scale of milliseconds [143, 144] is shown in Fig. 16 [145]. GRB afterglows can extend for days [143].

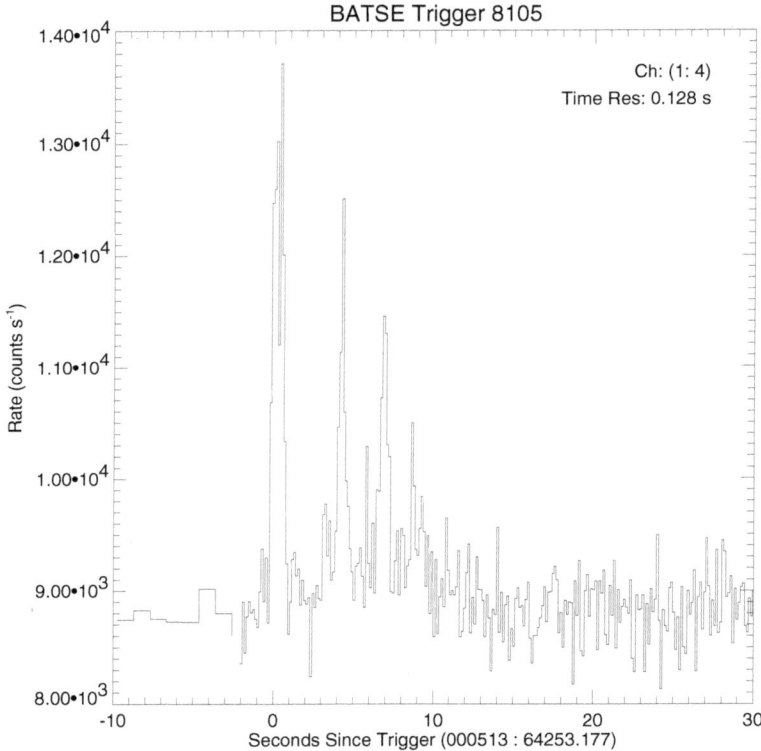

FIG. 16: An example of the temporal structure of a GRB as recorded by BATSE [145]. Note the two time scales: a duration of several seconds and a fluctuation time scale of a fraction of a second.

2. *A Brief History of Gamma Ray Bursts*

Gamma Ray Bursts (GRB) were accidentally discovered in the late 1960's by the military Vela satellites, intended to monitor nuclear tests in space forbidden by the Outer Space

Treaty between the United States and the Soviet Union [146]. The Vela observation of a short, intense burst of MeV gamma rays was originally considered to be a possible signal from an advanced extra-terrestrial civilization . The idea was quickly reconsidered. In 1973, the discovery was announced to the public [146]. Shortly after, the observation was verified by the Soviet IMP-6 satellite [147].

Until the 1990's, the high intensity of GRB led astronomers to the belief that they were galactic in origin. In 1991, the BATSE (Burst And Transient Satellite Experiment) detector on the Compton Gamma Ray Observatory was launched. BATSE observed roughly one burst per day within its field of view of about one third of the sky. The observations showed total isotropy of GRB over the entire sky, thus ruling out galactic origin [148]. The cosmological origin of GRB implies that they release up to a solar mass of energy, in seconds time. Their cosmological origin was subsequently confirmed by afterglow observations, first made in 1997 by the Beppo-SAX satellite [140]. Afterglow observations were made in X-ray, optical and longer wavelengths with an angular resolution of arc-minute precision and with measurement of the redshift. To date, dozens of GRB afterglows have been observed, nearly all of which have resulted in the identification of the host galaxy [149, 150, 151].

Although progress has been made in our understanding of GRB, many questions remain unanswered. Most importantly, the progenitor(s) of GRB remain an open question. In the next section, we describe some of the most likely candidates.

3. GRB Progenitors?

The observed characteristics of GRB require an original event with a large amount of energy ($\sim M_\odot$) in a very compact volume ($R_0 \sim 100\,\mathrm{km}$). The phenomenology that describes observations is that of a fireball expanding with highly relativistic velocity, powered by radiation pressure. The nature of the "inner engine" that initiates the fireball remains an open question. Afterglow observations have recently shown that GRB are predominantly generated in host galaxies and are likely the result of a stellar process. Research into a variety of stellar progenitors has been pursued.

The "Collapsar" scenario, where a super-massive star undergoes core collapse resulting in a failed supernova, is one of the most common models proposed for the fireball's inner engine [152, 153]. As matter falls into the black hole created in this process, gravitational energy is transfered to bulk kinetic energy and the fireball is generated.

The strength of magnetic fields and the angular momentum of the stellar object(s) involved can play an important role in the dynamics of the core collapse process. For example, "Magnetars" are a subset of the core collapse model which result in a rapidly spinning neutron star with an extremely strong magnetic field [154, 155, 156]. Objects with sufficient angular momentum can undergo a "supranova" process where their core collapse takes place in two stages, possibly separated by months or years [157, 158]. In this scenario, the object's large angular momentum prevents a fraction of the matter from falling into the fireball initially.

Compact objects in close binary orbits are also likely candidates for fireball progenitors. The "hypernovae" scenario is similar to the core collapse models, but includes a secondary stellar object in the dynamics [159, 160, 161, 162]. Similarly, neutron star binaries or neutron star-black hole binaries (or possibly white dwarf – neutron star or black hole binaries) which lose sufficient angular momentum through gravitation radiation can undergo a merger. Such a merger is expected to generate a black hole surrounded by debris. As this debris is accreted into the black hole, the required fireball is generated [161, 163, 164, 165, 166, 167, 168].

Finally, if primordial strange hadrons exist, a "seed" of strange matter may start a chain reaction converting a neutron star into a strange star made entirely of strange matter [169, 170, 171]. This conversion would release the majority of the star's binding energy as it contracts, thus generating a compact fireball similar to that required for GRB dynamics.

Recent evidence indicates the presence of emission lines in GRB [172]. This evidence strengthens the argument for progenitors involving collapsing stars.

The problem of GRB progenitors is likely to have an experimental solution. Possible progenitor-specific signatures may be found by using gravitational waves [173] or neutrinos as astronomical probes, or by more detailed study of afterglows [174, 175].

It is interesting to note that the bimodal distribution of GRB durations may be an indication of multiple GRB classes and associated progenitors.

4. Fireball Dynamics

a. The Fireball The dynamics of a gamma ray burst fireball is similar to the physics of the early universe. Initially, there is a radiation dominated soup of leptons and photons and few baryons. This perfect fluid has the equation of state $P = \rho/3$ and is initially hot enough to freely produce electron-positron pairs. The luminosity of a burst can be related to the number density of photons n_γ:

$$L = 4\pi R_0^2 c n_\gamma E_\gamma, \tag{13}$$

where R_0 is the initial radius of the source, i.e. prior to expansion. The optical depth of a photon before pair production is determined by the photon density and the interaction cross section [176]:

$$\tau_{\text{opt}} = \frac{R_0}{\lambda_{\text{int}}} = R_0 n_\gamma \sigma_{\text{Th}} = \frac{L \sigma_{\text{Th}}}{4\pi R_0 c E_\gamma} \sim 10^{15} \left(\frac{L_\gamma}{10^{52}\text{erg/s}}\right)\left(\frac{100\,\text{km}}{R_0}\right)\left(\frac{1\,\text{MeV}}{E_\gamma}\right). \tag{14}$$

Here λ_{int} is the interaction length of a photon as a result of pair production and Thomson scattering. These cross sections are roughly equal with the Thomson cross section $\sigma_{\text{Th}} \simeq 10^{-24}\text{cm}^2$.

With an optical depth of order $\sim 10^{15}$, photons are trapped in the fireball. This results in the highly relativistic expansion of the fireball powered by radiation pressure [168, 177]. The fireball will expand with increasing velocity until it becomes transparent and the radiation is released. This results in the visual display of the GRB. By this time, the expansion velocity has reached highly relativistic values of order $\gamma \simeq 300$.

Besides leptons and photons, the fireball contains some baryons. During expansion, the opaque fireball cannot radiate and any nucleons present are accelerated as radiation is converted into bulk kinetic energy. When the radiation is emitted, there is a transition from radiation to matter dominance of the fireball. At this stage, the radiation pressure is no longer important and the expanding fireball coasts without acceleration. The expansion velocity remains constant with $\gamma \simeq \eta \equiv \frac{L}{Mc^2}$ that is determined by the amount of baryonic matter present, often referred to as the baryon loading [178, 179]. The phenomenology will reveal values of η between 10^2 and 10^3 [136, 176, 180]. The formidable appearance of the GRB display is simply associated with the large boost between the fireball and the observer who detects highly boosted energies and contracted times.

The exploding fireballs original size, R_0, is that of the compact progenitor, for instance the black hole created by the collapse of a massive star. As the fireball expands the flow is shocked in ways familiar from the emission of jets by the black holes at the centers of active galaxies or mini-quasars. (A way to visualize the formation of shocks is to imagine that

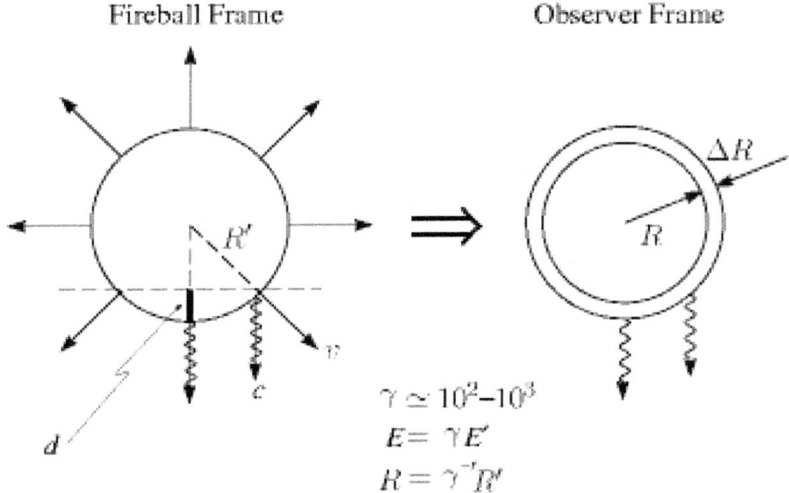

FIG. 17: Diagram of GRB fireball kinematics assuming no beaming. Primed quantities refer to the comoving frame. Unprimed quantities refer to the observer's frame.

infalling material accumulates and chokes the black hole. At this point a blob of plasma is ejected. Between these ejections the emission is reduced.) The net result is that the expanding fireball is made up of multiple shocks. These are the sites of the acceleration of particles to high-energy and the seeds for the complex millisecond structures observed in individual bursts; see Fig. 16. (Note that these shocks expand with a range of velocities and they will therefore collide providing a mechanism to accelerate particles to high-energy.) The characteristic width of these shocks in the fireball frame is $\delta R' = \gamma c \Delta t$, where $\Delta t \simeq 0.01$ sec. For an alternative scenario; see Ref. [181, 182, 183].

An expanding shock is seen by the observer as an expanding shell of thickness $c\Delta t = R_0$ and radius R; see Fig. 17. Here Δt is the time scale of fluctuations in the burst fireball; it is related to R' by:

$$R' = \gamma^2 c \Delta t = \gamma^2 R_0, \tag{15}$$

with primed quantities referring from now on to the frame where the fireball is at rest. Two, rather than a single γ-factor, relate the two quantities because of the geometry that relates the radius R' to the time difference between photons emitted from a shell expanding with a velocity v; see Fig. 17 [137]. Introducing the separation d of the two photons along the line of sight, we note that

$$\Delta t \sim \frac{d}{c} \sim \frac{1}{c}(R' - R'\frac{v}{c}) \simeq \frac{R'}{2c}(1 - \frac{v^2}{c^2}) \sim \frac{R'}{c\gamma^2}, \tag{16}$$

using the relativistic approximation that $1 + \frac{v^2}{c^2} \simeq 1$.

We next calculate the energy of the burst. In the observer frame

$$E = U \times V \propto T^4 \times R^2 \Delta R \propto T^4 R^2 \propto \gamma^4 T'^4 R^2, \tag{17}$$

where U is the energy density and V is the fireball shell volume. In the fireball frame,

$$E' = U' \times V' \propto T'^4 R'^3 \propto \gamma^6 T'^4. \tag{18}$$

Energy conservation requires that E and E' remain constant during expansion of the fireball. In the fireball frame, this results in the usual blackbody relation that T'^4 is proportional to R'^{-3} or, using Eq. 18, proportional to γ^{-6}. Substituting into the expression for E, we obtain that

$$E \propto \gamma^{-2} R^2, \tag{19}$$

or, because E is constant, that

$$R \propto \gamma. \tag{20}$$

Thus we obtain the important result that, with expansion, the γ-factor grows linearly with R until reaching the maximum value η.

b. *The Observed GRB Spectrum: Synchroton and Inverse Compton Scattering* The broken power-law gamma ray spectrum of GRB, with two distinct spectral slopes, is far from a blackbody spectrum. The observed spectrum, therefore, clearly indicates that fireball photons do not sufficiently interact to thermalize prior to escaping the fireball. After escaping, the photons show spectral features characteristic of the high-energy, non-thermal emission by supernova remnants and active galaxies. Here photons up to MeV-energy can be produced by synchrotron radiation, with some reaching, possibly, up to TeV energies by inverse Compton scattering on accelerated electrons [34, 136]. These processes have been modeled for the expanding fireball and successfully accommodate observed GRB spectra. This represents a major success of the relativistic fireball phenomenology [184, 185, 186, 187, 188].

To produce the non-thermal spectrum, special conditions must prevail [136, 137, 176, 180]. The photons must not thermalize prior to the time when the shock becomes transparent and the observed radiation released. Conversely, if they decouple too early, there is insufficient time for the synchotron and inverse Compton scattering processes to produce the observed spectrum. This requires that the expansion time of the shockwave and the time λ_{int}/c for photons and electrons to interact by Thompson scattering be similar:

$$\frac{R}{\gamma c} \sim (n_e c \sigma_T)^{-1}. \tag{21}$$

Here σ_T is the Thompson cross section and n_e is the electron number density in the fireball. The latter can be related to the mass flux $\dot{M}$ with the assumption that $n_e \cong n_p$:

$$\dot{M} = 4\pi R^2 c \gamma n_p m_p. \tag{22}$$

As required by mass conservation, $\dot{M}$ is independent of R since $n_p \propto \gamma^{-3}$ and $R \propto \gamma$. In terms of luminosity,

$$n_e \cong \frac{\dot{M}}{4\pi R^2 c \gamma m_p} = \frac{L}{\gamma \eta c^3 4\pi R^2 m_p}, \tag{23}$$

where $\eta = L/\dot{M}c^2$ is the ratio of luminosity to mass previously introduced. η is also referred to as the dimensionless entropy and, as previously derived, $\gamma \simeq \eta$ after expansion of the fireball. We can rewrite the condition of Eq. 21 for producing the observed non-thermal spectrum as

$$\frac{R n_e \sigma_T}{\gamma} = \frac{L \sigma_T}{\gamma^2 \eta c^3 4\pi R m_p} = \frac{1}{\gamma^3} \frac{\eta^{*4}}{\eta} \simeq 1, \tag{24}$$

where the critical dimensionless entropy η^* is defined as

$$\eta^* \equiv \left(\frac{L \sigma_T}{c^3 4\pi R_0 m_p} \right)^{1/4} \simeq 1000 \times \left(\frac{L}{10^{52} \text{erg/s}} \right)^{1/4} \left(\frac{100 \text{ km}}{R_0} \right)^{1/4}. \tag{25}$$

At decoupling, $\eta = \gamma$ and Eq. 24 is satisfied, provided $\eta = \eta^*$. The condition for the fireball to produce the correct non-thermal, synchrotron/inverse Compton spectrum is realized with the expansion time matching the Thompson scattering time. For values of η that are significantly larger (smaller), the decoupling of the radiation will occur too early (late) thus limiting η, as well as the final value of γ, to the range of $10^2 - 10^3$.

c. Jets and Beaming Observations imply that the total amount of energy emitted in gamma rays by a GRB are typically in the range of $10^{52} - 10^{54}$ ergs, i.e. a large fraction of a solar mass. For some bursts, it may exceed a solar mass. This, as well as the difficulty of converting such an unusually large fraction of primary energy into gamma rays, strongly suggests that GRB are beamed. Beaming reduces the total amount of energy by a factor of $\Omega/4\pi$, where Ω is the solid angle in which the observed gamma rays are emitted. Most proposed progenitors naturally predict a rotating stellar source that is likely to produce beamed emission. Relativistic beaming is possible down to an angular size of $\Omega > \gamma^{-2}$, although larger angles are of course possible [189, 190, 191].

In the presence of beaming, the number of bursts is increased by a factor of $4\pi/\Omega$ in order to account for bursts that do not point towards earth and are, therefore, not observed. For typical Lorentz factors of $\gamma \sim 300$ and a minimum beaming angle of $\Omega > \gamma^{-2} \sim 10^{-5}$, on the order of one GRB per galaxy per year is required to accommodate the observations [190]. It is important to note that most of the diffuse neutrino fluxes calculated in this review are independent of beaming because the reduced energy for a single burst is compensated by their increased frequency.

5. *Ultra High-energy Protons From GRB?*

As previously discussed, it may be possible to accelerate protons to energies above 10^{20} eV in GRB shocks [180, 192, 193]. GRB within the GZK radius of 50-100 Mpc, could therefore be the source of the ultra high-energy cosmic rays (UHECR's) [96, 97, 180, 192, 193, 194, 195, 196, 197, 198]. To accelerate protons to this energy, several conditions have to be satisfied. First, the acceleration time $t_a \sim AR_L/c$, where A is a factor of order 1 and $R_L = E/eB\gamma$ is the Larmor radius, must not exceed the duration of the burst $R/\gamma c$,

$$\frac{AE}{\gamma eB} \lesssim \frac{R}{\gamma}, \tag{26}$$

or

$$B \gtrsim \frac{AE}{eR} \simeq A \times 10\,\text{tesla}\left(\frac{E}{10^{20}\text{eV}}\right)\left(\frac{10^{11}\text{m}}{R}\right). \tag{27}$$

Second, energy losses due to synchrotron radiation must not exceed the energy gained by acceleration. The synchrotron loss time is given by

$$t_{syn} = \frac{\lambda_{int}}{c} = \frac{1}{cn_e \sigma_T}. \quad (28)$$

The number density of electrons (in the rest frame) is given by

$$n_e = \frac{m_e^2 c^4 B^2}{6\pi}. \quad (29)$$

Therefore,

$$t_{syn} = \frac{6\pi}{\sigma_T m_e^2 c^5 B^2 \gamma_p} = \frac{6\pi m_p^4 c^3}{\sigma_T m_e^2 E B^2}. \quad (30)$$

For synchrotron energy losses to be less than the energy gained by acceleration,

$$\frac{6\pi m_p^4 c^3}{\sigma_T m_e^2 E B^2} \gtrsim A \times \frac{E}{\gamma e c B}, \quad (31)$$

or

$$B \lesssim \frac{1}{A} \times 10\,\text{tesla} \left(\frac{\gamma}{300}\right)^2 \left(\frac{10^{20}\text{eV}}{E}\right)^2. \quad (32)$$

Combining above requirements, we get

$$A \times 10\left(\frac{E}{10^{20}\text{eV}}\right)\left(\frac{10^{11}\text{m}}{R}\right)\text{tesla} \lesssim B \lesssim \frac{1}{A} \times 10\,\text{tesla}\left(\frac{\gamma}{300}\right)^2\left(\frac{10^{20}\text{eV}}{E}\right)^2, \quad (33)$$

or

$$R \gtrsim A^2 \times 10^{10}\,\text{meters}\left(\frac{300}{\gamma}\right)^2\left(\frac{E}{10^{20}\text{eV}}\right)^3. \quad (34)$$

From simple fireball kinematics, we previously derived that

$$R \lesssim \gamma^2 c \Delta t, \quad (35)$$

where Δt is ~ 10 msec. Combining this with Eq. 34 leads to the final requirement:

$$A^2 \times 10^{10}\left(\frac{300}{\gamma}\right)^2\left(\frac{E}{10^{20}\text{eV}}\right)^3 \lesssim R \lesssim \gamma^2 c \Delta t, \quad (36)$$

or

$$\gamma \gtrsim A^{1/2} \times 130 \left(\frac{E}{10^{20}\text{eV}}\right)^{3/4}\left(\frac{.01\text{sec}}{\Delta t}\right)^{1/4}, \quad (37)$$

which can indeed be satisfied for the values of $\gamma = 10^2 - 10^3$ previously derived from fireball phenomenology. We conclude that bursts with Lorentz factors $\gtrsim 100$ can accelerate protons to $\sim 10^{20}$ eV. The long acceleration time of 10-100 seconds implies however that the fireball extends to a large radius where surrounding matter may play an important role in the kinematics of the expanding shell.

Finally, it can be shown that proton energy losses from $p-\gamma$ interactions will not interfere with acceleration to high-energy. These will, in fact, be the source of high-energy neutrinos associated with the beam of high-energy protons. We will discuss this further on.

6. Neutrino Production in GRB: the Many Opportunities

Several mechanisms have been proposed for the production of neutrinos in GRB. We summarize them first:

- **Thermal Neutrinos: MeV Neutrinos**

  As with supernovae, GRB are expected to radiate the vast majority of their initial energy as thermal neutrinos. Although the details are complex and are likely to depend on the progenitor, a neutrino spectrum with a higher temperature than a supernova may be expected. Observation is difficult because of the great distances to GRB, although we should keep in mind that a nearby GRB may not be less frequent than a galactic supernova; see the section on beaming.

- **Shocked Protons: TeV-EeV Neutrinos**

  Protons accelerated in GRB can interact with fireball gamma rays and produce pions that decay into neutrinos. While astronomical observations provide information on the fireball gamma rays, the proton flux is a matter of speculation. A definite neutrino flux is however predicted when assuming that GRB produce the highest energy cosmic rays.

- **Decoupled Neutrons: GeV Neutrinos**

  In a GRB fireball, neutrons can decouple from protons in the expanding fireball. If their relative velocity is sufficiently high, their interactions will be the source of pions and, therefore, neutrinos. Typical energies of the neutrinos produced are much lower than those resulting from interactions with gamma rays.

7. Thermal MeV Neutrinos from GRB

As is the case for a supernova, we expect that in GRB, thermal neutrinos are produced escaping with the majority of the total energy [199, 200]. The dynamics are somewhat different, however. The temperature of the photons in a supernova is of order 10 MeV as derived from the familiar estimate:

$$U_\gamma = \frac{4\sigma T^4}{c} \to T_\gamma \simeq \left(\frac{E_{\text{in }\gamma\text{'s}}}{V}\frac{c}{4\sigma}\right)^{1/4} \simeq 11\,\text{MeV} \left(\frac{E_{\text{in }\gamma\text{'s}}}{10^{52}\text{ergs}}\right)^{1/4}\left(\frac{100\,\text{km}}{R}\right)^{3/4}, \tag{38}$$

where $\sigma \simeq 5.67 \times 10^{-8}\,\text{km/sec}^3$ is the Stefan-Boltzman constant and U_γ is the energy density of the initial plasma. We copy this estimate for the neutrinos produced in the pre-fireball phase of a GRB taking into account the much larger energy emitted, typically 2 orders of magnitude:

$$T_\nu = T_\gamma \left(\frac{E_{\text{in }\nu\text{'s}}/E_{\text{in }\gamma\text{'s}}}{h_\nu/h_\gamma}\right)^{1/4} \simeq 28\,\text{MeV}\left(\frac{E_{\text{in }\nu\text{'s}}}{10^{54}\text{ergs}}\right)^{1/4}\left(\frac{100\,\text{km}}{R}\right)^{3/4}. \tag{39}$$

Here $h_\nu = 2 \times 3 \times 7/8 = 21/4$ and $h_\gamma = 2$ are the degrees of freedom available to each particle type. This yields neutrinos with average energy:

$$E_{\nu,\text{ave}} \simeq 3.15\,T_\nu \simeq 90\,\text{MeV}\left(\frac{E_{\text{in }\nu\text{'s}}}{10^{54}\text{ergs}}\right)^{1/4}\left(\frac{100\,\text{km}}{R}\right)^{3/4}. \tag{40}$$

The flux of neutrinos can now be calculated from the average energy of roughly 100 MeV per individual neutrino and the total energy available:

$$N_\nu \simeq \frac{E_{\text{in }\nu\text{'s}}}{E_{\nu,\text{ave}}}\frac{1}{4\pi D^2} \simeq 6 \times 10^{10}\,\text{km}^{-2}\left(\frac{E_{\text{in }\nu\text{'s}}}{10^{54}\text{ergs}}\right)^{3/4}\left(\frac{R}{100\,\text{km}}\right)^{3/4}\left(\frac{3000\,\text{Mpc}}{D}\right)^2. \tag{41}$$

Neutrinos with this energy are below threshold for the detection methods previously described. For supernova 1987A, predominantly electron anti-neutrinos were observed by the electromagnetic showers generated inside the detector by the process $\bar{\nu}_e + p \to n + e^+$. Underground detectors are too small to detect the above flux because of the cosmological distance to the source. In large under-ocean detectors, the signal is drowned in the $\sim 50\,\text{kHz}$ noise in the photomultipliers from potassium decay. Only a large Cerenkov detector embedded in sterile ice can possibly detect GRB but even here the signal-to-noise is marginal unless the source is within our local cluster. Detailed calculations have been performed in Ref. [199].

8. Shocked Protons: PeV Neutrinos

Assuming that GRB are the sources of the highest energy cosmic rays and that the efficiency for conversion of fireball energy into the kinetic energy of protons is similar to that for electrons, the production of PeV neutrinos is a robust prediction of the relativistic fireball model [94, 137, 194]. Neutrinos are produced in interactions of accelerated protons with fireball photons, predominantly via the processes

$$p\gamma \to \Delta \to n\pi^{\pm} \tag{42}$$

and

$$p\gamma \to \Delta \to p\pi^0 \tag{43}$$

which have very large cross sections of 10^{-28}cm^2. The charged π's subsequently decay producing charged leptons and neutrinos, while neutral π's may generate high-energy photons observable in TeV energy air Cerenkov detectors.

For the center-of-mass energy of a proton-photon interaction to exceed the threshold energy for producing the Δ-resonance, the comoving proton energy must exceed

$$E'_p > \frac{m_\Delta^2 - m_p^2}{4E'_\gamma}. \tag{44}$$

Therefore, in the observer's frame,

$$E_p > 1.4 \times 10^{16} \text{eV} \left(\frac{\gamma}{300}\right)^2 \left(\frac{1\text{MeV}}{E_\gamma}\right), \tag{45}$$

resulting in a neutrino energy

$$E_\nu = \frac{1}{4}\langle x_{p\to\pi}\rangle E_p > 7 \times 10^{14} \text{eV} \left(\frac{\gamma}{300}\right)^2 \left(\frac{1\text{MeV}}{E_\gamma}\right). \tag{46}$$

Here $\langle x_{p\to\pi}\rangle \simeq .2$ is the average fraction of energy transferred from the initial proton to the produced pion. The factor of $1/4$ is based on the estimate that the 4 final state leptons in the decay chain $\pi^{\pm} \to \bar\nu_\mu \mu \to \bar\nu_\mu e \nu_e \bar\nu_e$ equally share the pion energy.

We already discussed the fireballs expansion and the formation of shocks that are the sites of the acceleration of particles to high-energy and the seeds for the complex millisecond structures observed in individual bursts. The characteristic width of a shock in the fireball frame is $\Delta R' = \gamma c \Delta t$, where $\Delta t \simeq 0.01$ sec.

As the kinetic energy in fireball protons increases with expansion, a fraction of this energy is converted into pions once the protons are accelerated above threshold for pion production. The fraction of energy converted to pions is estimated from the number of proton interactions occurring within a shock of characteristic size $\Delta R'$:

$$f_\pi \simeq \frac{\Delta R'}{\lambda_{p\gamma}} \langle x_{p\to\pi}\rangle. \tag{47}$$

The proton interaction length $\lambda_{p\gamma}$ in the photon fireball is given by

$$\frac{1}{\lambda_{p\gamma}} = n_\gamma \sigma_\Delta. \tag{48}$$

Here n_γ is the number density of photons in the fireball frame and $\sigma_\Delta \sim 10^{-28}\text{cm}^2$ is the proton-photon cross section at the Δ-resonance. The photon number density is the ratio of the photon energy density and the photon energy in the comoving frame:

$$n_\gamma = \frac{U'_\gamma}{E'_\gamma} = \left(\frac{L_\gamma \Delta t/\gamma}{4\pi R'^2 \Delta R'}\right) \bigg/ \left(\frac{E_\gamma}{\gamma}\right). \tag{49}$$

Using the fireball kinematics of Eqs. 15 and 16

$$n_\gamma = \left(\frac{L_\gamma}{4\pi c^3 \Delta t^2 \gamma^6}\right) \bigg/ \left(\frac{E_\gamma}{\gamma}\right) = \frac{L_\gamma}{4\pi c^3 \Delta t^2 \gamma^5 E_\gamma}. \tag{50}$$

Thus we obtain the fraction of proton energy converted to π's in the expansion:

$$f_\pi \simeq \frac{L_\gamma}{E_\gamma} \frac{1}{\gamma^4 \Delta t} \frac{\sigma_\Delta \langle x_{p\to\pi}\rangle}{4\pi c^2} \simeq .13 \times \left(\frac{L_\gamma}{10^{52}\text{erg/s}}\right)\left(\frac{1\text{MeV}}{E_\gamma}\right)\left(\frac{300}{\gamma}\right)^4\left(\frac{.01\text{sec}}{\Delta t}\right). \tag{51}$$

For $L \sim 10^{52}$ erg/sec, $\Delta t \sim 10$ msec and $\gamma \simeq 300$, this fraction is on the order of 10 percent. This quantity strongly depends on the Lorentz factor γ. Even modest burst-to-burst fluctuations in γ around the average value of 300 can result in a PeV neutrino flux dominated by a few bright bursts; we will return to a discussion of fluctuations further on [201, 202].

In order to normalize the neutrino flux we introduce the assumption that GRB are the source of cosmic rays above the ankle of the cosmic ray spectrum near $\sim 3 \times 10^{18}$ eV [96, 180, 192, 193, 194, 195, 196, 197, 198]. The flux in neutrinos can then be simply obtained from the total energy injected into cosmic rays and the average energy of a single neutrino [94]:

$$\phi_\nu \simeq \frac{c}{4\pi}\frac{U'_\nu}{E'_\nu} = \frac{c}{4\pi}\frac{U_\nu}{E_\nu} = \frac{c}{4\pi}\frac{1}{E_\nu}\left(\frac{1}{2}f_\pi t_H \frac{dE}{dt}\right), \tag{52}$$

or

$$\phi_\nu = 2 \times 10^{-14} \text{cm}^{-2}\text{s}^{-1}\text{sr}^{-1} \left(\frac{7 \times 10^{14}\text{eV}}{E_\nu}\right) \left(\frac{f_\pi}{0.125}\right) \left(\frac{t_H}{10\text{Gyr}}\right) \left(\frac{dE/dt}{4 \times 10^{44}\,\text{Mpc}^{-3}\text{yr}^{-1}}\right), \quad (53)$$

where $t_H \sim 10$ Gyrs is the Hubble time and $dE/dt \sim 4 \times 10^{44}$ ergs Mpc^{-3} yr^{-1} is the injection rate of energy into the universe in the form of cosmic rays above the ankle.

PeV neutrinos are detected by observing a charged lepton produced in the charged current interaction of a neutrino near the detector. For instance, the probability to detect a muon neutrino within a Cerenkov neutrino telescope's effective area is given by Eq. 11. At TeV-PeV energies the function $P_{\nu \to \mu}$ can be approximated by

$$P_{\nu \to \mu} \simeq 1.7 \times 10^{-6} E_{\nu,obs}^{0.8} (\text{TeV}), \quad (54)$$

where $E_{\nu,obs} = E_\nu/(1+z)$ is the observed neutrino energy. The rate of detected events is the convolution of the flux with the probability of detecting the neutrino

$$N_{events} = \int_{E_{thresh}}^{E_\nu^{max}} \phi_\nu P_{\nu \to \mu} \frac{dE_\nu}{E_\nu} \simeq 25 \text{ km}^{-2}\text{yr}^{-1} \quad (55)$$

This rate is significantly enhanced when fluctuations in distance, energy and (possibly) Lorentz factor are considered. The event rate is likely to be dominated by a few bright bursts rather than by a diffuse flux.

With the ability to look for GRB neutrino events in coincidence with gamma ray observations, i.e. in short time windows over which very little background accumulates, there is effectively no background for this neutrino signature of GRB.

9. Stellar Core Collapse: Early TeV Neutrinos

The core collapse of massive stars is, arguably, the most promising mechanism for generating GRB. The fireball produced is likely to be beamed in jets along the collapsed object's rotation axis. The mechanism is familiar from observations of jets associated with the central black hole in active galaxies. The jets subsequently run into the stellar matter accreting onto the black hole. If the jets successfully emerge from the stellar envelope a GRB results. Interestingly, failed "invisible" jets which do not emerge will not produce a GRB display but will still produce observable neutrinos [203, 205].

A beamed GRB jet expanding with a Lorentz factor $\gamma_{\text{jet}} \sim 100 - 1000$ through the stellar envelope, will be slowed down resulting in a smaller Lorentz factor at its leading edge

$\gamma_f \ll \gamma_{jet}$. Therefore, the fast particles in the tail will catch up with the slow particles in the leading edge and collide with a Lorentz factor $\gamma \approx \gamma_{jet}/2\gamma_f$ by simple addition of relativistic velocities. Once the jet emerges from the infalling stellar matter around $R \sim 10^6$ km, the density drops to around $\sim 10^{-7}$g/cm^3. Matching the energy densities on either side of the shock front requires

$$\gamma_f^2 \times 10^{-7} \text{g/cm}^3 \simeq \left(\gamma_{jet}/2\gamma_f\right)^2 n_p m_p, \tag{56}$$

where n_p is the comoving proton number density in the jet. It is related to the luminosity of the burst; see Eq. 23:

$$n_p = \frac{L_{iso}}{4\pi R^2 \gamma_{jet}^2 m_p c^3}. \tag{57}$$

Here $L_{iso} = L\frac{\Omega}{4\pi}$ is the inferred from a non-beamed flux. Eqs. 56 and 57 determine the value of γ_f:

$$\gamma_f \simeq \left(\frac{L_{iso}}{16\pi \rho R^2 c^3}\right)^{1/4} \simeq 3 \times \left(\frac{L_{iso}}{10^{52} \text{erg/s}}\right)^{1/4} \left(\frac{10^{-7}\text{g/cm}^3}{\rho}\right)^{1/4} \left(\frac{10^7 \text{km}}{R}\right)^{1/2}. \tag{58}$$

With this low value of the Lorentz factor, the fireball remains opaque to gamma rays as described in the section on fireball dynamics. The radiation thermalizes with a temperature determined by its energy density U'_γ and the Stefan-Boltzman law $U'_\gamma = 4\sigma T'^4/c$. We find

$$T'_\gamma \simeq \left(\frac{4\gamma_f^2 \rho c^3}{4\sigma}\right)^{1/4} \simeq 2.2 \text{ keV} \left(\frac{\rho}{10^{-7}\text{g/cm}^3}\right)^{1/4} \left(\frac{\gamma_f}{3}\right)^{1/2}. \tag{59}$$

The rest of the calculation follows the previous section. Protons traveling through this thermal photon plasma produce pions, predominantly via the Δ-resonance, for energies

$$E'_p > \frac{m_\Delta^2 - m_p^2}{4E'_\gamma} \simeq 7 \times 10^4 \text{ GeV} \left(\frac{2.2 \text{ keV}}{E'_\gamma}\right). \tag{60}$$

Primed energies refer to the comoving frame with $\gamma_f E' = E$. Neutrinos emerge from the interactions with energy

$$E_\nu = \frac{1}{4}\langle x_{p\to\pi}\rangle E_p \simeq 10 \text{ TeV} \left(\frac{\gamma_f}{3}\right)\left(\frac{2.2 \text{ keV}}{T'_\gamma}\right). \tag{61}$$

For mildly relativisitic conditions, the fraction of protons converted to pions is expected to be high in the very dense plasma, i.e. of order unity. The neutrino flux observed from a single GRB at a distance D is calculated from the total energy emitted in neutrinos and the average energy of a single neutrino:

$$\phi_\nu \simeq \frac{E_{\text{iso}}\langle x_{p\to\pi}\rangle}{16\pi D^2 E_\nu} \simeq .003\,\nu's\,m^{-2} \times \left(\frac{E_{\text{iso}}}{10^{53}\text{ergs}}\right)\left(\frac{10\text{TeV}}{E_\nu}\right)\left(\frac{3000\text{Mpc}}{D}\right)^2. \qquad (62)$$

This flux of TeV neutrinos from a single burst results in

$$N_{events} \sim \phi_\nu \times P_{\nu\to\mu} \sim .05\left(\frac{E_{\text{iso}}}{10^{53}\text{ergs}}\right)\left(\frac{3000\text{Mpc}}{D}\right)^2 \qquad (63)$$

events observed per year, per square kilometer of the detector. Here $E_{\nu,obs} = E_\nu/(1+z)$ is the observed neutrino energy and, for TeV-energy neutrinos, we used the approximation

$$P_{\nu\to\mu} \simeq 1.3 \times 10^{-6} E_{\nu,obs}(\text{TeV}). \qquad (64)$$

The event rate is low. Bursts within a few hundred megaparsecs (~ 10 bursts per year as well as an additional unknown number of "invisible" bursts from failed GRB) may produce multiple TeV neutrino events in a kilometer scale detector. This signature is unique to supernova progenitors.

10. UHE Protons From GRB: EeV Neutrinos

Recent observations of GRB afterglows show evidence that GRB explode into an interstellar medium, consistent with the speculations that they are collapsing or merging stars. Shocks will be produced when the GRB runs into the interstellar medium, including a reverse shock that propagates back into the burst ejecta. Electrons and positrons in the reverse shock radiate an afterglow of eV-keV photons that represent a target for neutrino production by ultra high-energy protons accelerated in the burst [204].

The fraction of proton energy going into π-production is calculated as before following Eq. 47,

$$f_\pi \simeq \frac{\Delta R'}{\lambda_{p\gamma}}\langle x_{p\to\pi}\rangle, \qquad (65)$$

$$f_\pi \simeq \frac{L_\gamma(E_{\gamma,min})}{E_{\gamma,min}}\frac{1}{\gamma_{rs}^4 \Delta t}\frac{\sigma_\Delta \langle x_{p\to\pi}\rangle}{4\pi c^2}, \qquad (66)$$

where γ_{rs} is the Lorentz factor of the reverse shock. For the afterglow, the relevant time scale is 10-100 seconds and the luminosity is $L_\gamma \propto E_\gamma^{-1/2}$ [176]. $E_{\gamma,min}$, the minimum photon energy to produce pions via the Δ-resonance, is given by:

$$E_{\gamma,min} = \frac{\gamma_{rs}^2(m_\Delta^2 - m_p^2)}{4E_p}. \tag{67}$$

Therefore 10^{21} eV protons can kinematically produce π's on photons with energy as low as 10 eV. Combining Eq. 66 and Eq. 67 we find

$$f_\pi \simeq .003 \times \left(\frac{10\,\mathrm{eV}}{E_{\gamma,min}}\right)\left(\frac{E_p}{10^{21}\mathrm{eV}}\right)^{1/2}\left(\frac{300}{\gamma_{rs}}\right)^5\left(\frac{20\mathrm{sec}}{\Delta t}\right). \tag{68}$$

Note that above keV energy, the photon luminosity follows the broken spectrum with $L_\gamma \propto E_\gamma^{-1}$ and, therefore, $f_\pi \propto E_p$ rather than $f_\pi \propto E_p^{1/2}$.

Associating the accelerated beam with the observed ultra high-energy cosmic ray flux, $dN_p/dE_p \sim AE_p^{-2}$, where $A \sim 5 \times 10^{-4} \mathrm{m}^{-2}\mathrm{s}^{-1}\mathrm{sr}^{-1}\mathrm{GeV}$, and using $E_\nu \sim .05 E_p$, see Eq. 46, the resulting neutrino flux is given by

$$\frac{dN_\nu}{dE_\nu}(E_\nu) \sim \frac{dN_p}{dE_p}(E_p = 20E_\nu) \times f_\pi(E_p = 20E_\nu), \tag{69}$$

$$\frac{dN_\nu}{dE_\nu}(E_\nu) \sim A \times (20E_\nu)^{-2} \times .003 \times \left(\frac{10\mathrm{eV}}{E_{\gamma,min}}\right)\left(\frac{20E_\nu}{10^{21}\mathrm{eV}}\right)^{1/2}\left(\frac{300}{\gamma_{rs}}\right)^5\left(\frac{20\mathrm{sec}}{\Delta t}\right), \tag{70}$$

$$\frac{dN_\nu}{dE_\nu}E_\nu^2(E_\nu) \sim 2 \times 10^{-14} E_\nu^{1/2}(\mathrm{GeV}) \times \left(\frac{10\mathrm{eV}}{E_{\gamma,min}}\right)\left(\frac{300}{\gamma_{rs}}\right)^5\left(\frac{20\mathrm{sec}}{\Delta t}\right)\mathrm{m}^{-2}\mathrm{s}^{-1}\mathrm{sr}^{-1}\mathrm{GeV}. \tag{71}$$

It is important to note that if the burst occurs in a region of higher density gas, as can be the case for a collapsing star, reverse shocks are produced earlier and, therefore, with smaller Lorentz factors. This results in $f_\pi \simeq 1$. Then,

$$\frac{dN_\nu}{dE_\nu}(E_\nu) \simeq \frac{dN_p}{dE_p}(E_p = 20E_\nu) \times 1 \simeq A \times (20E_\nu)^{-2}, \tag{72}$$

$$\frac{dN_\nu}{dE_\nu}E_\nu^2(E_\nu) \sim 10^{-6}\mathrm{m}^{-2}\mathrm{s}^{-1}\mathrm{sr}^{-1}\mathrm{GeV}. \tag{73}$$

In either case, the result is only valid above the threshold energy required to generates pions via the Δ-resonance,

$$E_\nu^{min} \simeq .05 E_p^{min} \simeq .05\frac{\gamma_{rs}^2(m_\Delta^2 - m_p^2)}{4E_\gamma^{max}} \sim 7 \times 10^{17}\mathrm{eV} \times \left(\frac{\gamma_{rs}}{300}\right)^2\left(\frac{1\mathrm{keV}}{E_\gamma^{max}}\right). \tag{74}$$

Below this threshold, ultra high-energy protons may still interact with non-thermal MeV photons, however.

The event rate in a neutrino telescope is calculated following Eq. 11. In the high-energy appoximation,

$$P_{\nu \to \mu} \simeq 1.2 \times 10^{-2} E_{\nu,obs}^{0.5}(\text{EeV}). \tag{75}$$

This yields

$$N_{events} \sim \int_{7\times 10^8 \,\text{GeV}}^{5\times 10^{10}\,\text{GeV}} 10^{-6} E_\nu^{-2} \times 3.7 \times 10^{-7} E_\nu^{0.5} dE_\nu \,\text{m}^{-2}\text{s}^{-1}\text{sr}^{-1} \sim .01\,\text{km}^{-2}\text{yr}^{-1}. \tag{76}$$

This is a very small rate indeed. The neutrino energy is, however, above the threshold for EeV telescopes using acoustic, radio or horizontal air shower detection techniques. This mechanism may represent an opportunity for detectors with very high threshold, but also large effective area to do GRB physics.

11. *The Decoupling of Neutrons: GeV Neutrinos*

The conversion of radiation into kinetic energy in the fireball will accelerate neutrons along with protons, especially if the progenitor involves neutron stars. Protons and neutrons are initially coupled by nuclear elastic scattering. If the expansion of the fireball is sufficiently rapid the neutrons and protons will no longer interact. Neutrons decouple from the fireball while protons are still accelerated. Protons and neutrinos may then achieve relative velocities sufficient to generate pions which decay into GeV neutrinos [206, 207]. We define the ratio of neutrons to protons as,

$$\xi \equiv \frac{n_n}{n_p}, \tag{77}$$

which initially remains constant during expansion. The fraction of neutrons which generate pions is calculated in the same way as in Eq. 47,

$$f_\pi \simeq \frac{\Delta R'}{\lambda_{pn}}. \tag{78}$$

We can relate the density of nucleons to the density of photons by the dimensionless entropy,

$$n_{p+n} \simeq n_\gamma \frac{\gamma E_\gamma}{\eta m_p c^2}. \tag{79}$$

Following the arguments used in our discussion of PeV neutrinos, we arrive at

$$f_\pi \simeq \frac{L_\gamma}{m_p \eta \gamma^3 \Delta t} \frac{\sigma_{np}}{4\pi c^2 (1+\xi)}, \tag{80}$$

where $\sigma_{np} \simeq 3 \times 10^{26}$ cm^2 is the neutron-proton cross section for pion production. As the neutrons and protons decouple, f_π approaches unity. Using the fact that γ asymptotically approaches η at the end of expansion, we see that decoupling occurs for

$$\eta \gtrsim \eta_{np} \equiv \left(\frac{L\sigma_{np}}{4\pi R_0 m_p c^3 (\xi+1)}\right)^{1/4} \simeq 400 \left(\frac{L}{10^{52} \text{erg/s}}\right)^{1/4} \left(\frac{100 \text{ km}}{R_0}\right)^{1/4} \left(\frac{2}{\xi+1}\right)^{1/4}. \quad (81)$$

In fact, the requirement for exceeding the threshold for π production is $\eta \geq 1.2\,\eta_{np}$ [206]. The scattering time is therefore longer than the expansion time by a factor $1.2^4 \simeq 2.1$ and $1 - e^{-2.1} \cong 88\%$ of the neutrons scatter. If threshold were exceeded by, say $\eta = 1.5\,\eta_{np}$, then more than 99% of the neutrons would scatter. It is therefore a reasonable approximation to assume that all neutrons produce π's as long as η is above threshold. The number of neutrons in the fireball is large with

$$N_n \simeq \frac{E}{m_p c^2} \frac{\xi}{1+\xi} \frac{1}{\eta} \sim 7 \times 10^{52} \left(\frac{E}{10^{53} \text{ergs}}\right) \left(\frac{2\xi}{1+\xi}\right) \left(\frac{500}{\eta}\right). \quad (82)$$

Above the pion threshold, every neutron interacts with a proton producing one of the following:

- $p + n \to p + p + \pi^- \to \bar{\nu}_\mu + \mu^- \to e^- + \bar{\nu}_e + \nu_\mu + \bar{\nu}_\mu$
- $p + n \to n + n + \pi^+ \to \nu_\mu + \mu^+ \to e^+ + \nu_e + \bar{\nu}_\mu + \nu_\mu$
- $p + n \to p + n + \pi^0 \to \gamma + \gamma$

Thus on average, each interaction produces two 30-50 MeV muon neutrinos and two 30-50 MeV electron neutrinos or two 70 MeV photons. The observed neutrino energy is

$$E_{\nu,obs} \simeq 30 - 50\,\text{MeV} \times \frac{\gamma}{1+z} \simeq 6 - 10\,\text{GeV} \times \left(\frac{\gamma}{400}\right)\left(\frac{2}{1+z}\right). \quad (83)$$

This energy is below the threshold of neutrino telescopes with the possible exception of Baikal and ANTARES provided it is built with a sufficiently dense arrangement of the photomultipliers.

$$P_{\nu \to \mu}(E_\nu \sim \text{GeV}) \sim 10^{-7} E_{\nu,\text{obs}}(\text{GeV}) \quad (84)$$

parametrizes the chance of detecting a $\sim$GeV muon neutrino in ANTARES as described by Eq. 11. This leads to an event rate from a single burst of

$$N_{events} \simeq N_\nu \frac{A_{\text{eff}}}{4\pi D^2} P_{\nu\to\mu} \tag{85}$$

where $N_\nu = 2N_n$ is the number of muon neutrinos emitted by the burst, A_{eff} is the detector's effective area and D is the distance to the burst. Positioning, for simplicity, all bursts at z=1, this reduces to

$$N_{events} \sim N_{bursts} \times \left(2 N_n \frac{A_{\text{eff}}}{4\pi D^2} P_{\nu\to\mu}(8\text{GeV})\right) \tag{86}$$

$$\sim .1 \times \left(\frac{N_{bursts}}{1000\text{yr}^{-1}}\right)\left(\frac{E}{10^{53}\text{ergs}}\right)\left(\frac{2\xi}{1+\xi}\right)\left(\frac{3-2\sqrt{2}}{2+z-2\sqrt{1+z}}\right)\left(\frac{A_{\text{eff}}}{.1\text{km}^2}\right)^{3/2} \text{yr}^{-1}. \tag{87}$$

This estimate is optimistic and the rate quite small. Observation requires a burst with favorable fluctuations in distance and energy. We discuss this important aspect of GRB detection next.

12. Burst-To-Burst Fluctuations and Neutrino Event Rates

We have focused so far on the diffuse flux of neutrinos produced collectively by all GRB. Given the possibility to observe individual GRB, it is important to consider the fluctuations that may occur in the dynamics from burst to burst. In the presence of large fluctuations, the relevant observable becomes the number of neutrinos observed from individual bursts [201, 202]. We will consider fluctuations in the distance to the burst D, and in its energy E. One can further contemplate fluctuations in the Lorentz factor, γ, or equivalently in η, the dimensionless entropy, although it is not clear whether energy and Lorentz factor are independent quantities.

The number of neutrinos from a single source at a distance D is proportional to D^{-2}. Considering burst-to-burst fluctuations in distance will enhance the fluxes previously calculated. This can be see as follows

$$\frac{\int_0^{ct_{\text{Hubble}}} dN(D)D^{-2}}{\int_0^{ct_{\text{Hubble}}} dN(D)D_{\text{ave}}^{-2}} = \frac{\int_0^{ct_{\text{Hubble}}} dD}{\int_0^{ct_{\text{Hubble}}} D^2 D_{\text{ave}}^{-2} dD} = \frac{D_{\text{ave}}^2 ct_{\text{Hubble}}}{\frac{1}{3}(ct_{\text{Hubble}})^3} = 3\frac{D_{\text{ave}}^2}{(ct_{\text{Hubble}})^2}, \tag{88}$$

where we have assumed an isotropic Euclidian distribution of sources. In contrast, the average distance of a burst D_{ave} is

$$\int_0^{D_{\text{ave}}} D^2 dD = \frac{1}{2}\int_0^{ct_{\text{Hubble}}} D^2 dD \to D_{\text{ave}} = 2^{-1/3} ct_{\text{Hubble}}. \tag{89}$$

Therefore, the event rate is enhanced by spatial fluctuations by a factor

$$3 \times 2^{-2/3} \simeq 1.9. \tag{90}$$

This enhancement factor should be applied to the diffuse flux. It also represents an increased probability to observe a single burst yielding multiple neutrino events. Assuming a more realistic cosmological distribution for starburst galaxies increases the effect of spatial fluctuations by an additional factor of 2 above the Euclidian result.

Next, we consider energy. Observations show that approximately one out of ten bursts is ten times more energetic than the median burst and approximately one out of a hundred is a hundred times more energetic. This results in an enhancement of the neutrino signal by a factor of

$$.89 \times E_{\text{median}} + .10 \times 10\, E_{\text{median}} + .01 \times 100\, E_{\text{median}} \simeq 2.9\, E_{\text{median}}. \tag{91}$$

If instead of a step function, a smooth distribution $\frac{dN}{dE} \propto E^{-2}$ covering two orders of magnitude is used, the enhancement factor is increased to 4.6.

We thus reach the very important conclusion that correct averaging of bursts over their cosmological spatial distribution and over their observed energy distribution enhances the neutrino signal by approximately one order of magnitude. Variations in distance and energy affect all GRB neutrino fluxes previously discussed, regardless of the production mechanism.

The third, and most important, quantity to vary from burst-to-burst is the Lorentz factor, or entropy. Its variation can greatly modify some fluxes, e.g. as $\sim \gamma^4$ for PeV neutrinos or $\sim \gamma^5$ for EeV neutrinos, and it barely affects others, e.g. GeV neutrinos produced by decoupled neutrons. The range over which γ can vary is, however, limited [136, 137, 176, 180]. The Lorentz factor i) must be roughly 300 to generate the observed non-thermal gamma ray spectrum, ii) must be large enough to produce the highest energy cosmic rays of $\sim 10^{20}$eV, and iii) the energy of shocked protons must be sufficient to be above threshold for producing pions on fireball gamma rays.

The last condition requires that

$$m_p \gamma \gtrsim \frac{m_\Delta^2 - m_p^2}{4 E_{\gamma,\text{obs}}} \Longrightarrow \gamma \gtrsim 170 \left(\frac{1\,\text{MeV}}{E_{\gamma,\text{obs}}}\right), \tag{92}$$

and the second that

$$\gamma \gtrsim A^{1/2} \times 130 \left(\frac{E}{10^{20}\text{eV}}\right)^{3/4} \left(\frac{.01\,\text{sec}}{\Delta t}\right)^{1/4} \tag{93}$$

following Eq. 37. Once again, A is a factor of order unity, E the maximum proton energy generated and Δt the time scale associated with the rapid variations observed in single bursts.

Together, these constraints limit any variation of the Lorentz factor to at most one order of magnitude. This does, however, correspond to a variation over four (five) orders of magnitude in the fraction of energy converted to π's which yield PeV (EeV) neutrinos. A change in Lorentz factor modifies the peak neutrino energy of the neutrinos as γ^2 and is, therefore, a secondary effect relative to the fraction of energy converted to pions.

How the Lorentz factor of GRB vary from burst-to-burst is an open question. It is interesting to note that even with conservative fluctuations of a factor of two or so (say between 200 and 400), the difference in neutrino flux exceeds one order of magnitude. Combined with variations in distance, and perhaps burst energy, the occasional nearby and bright (small γ, high E, or both) burst becomes a superior experimental signature to the diffuse flux. More detailed modeling has been presented in Ref. [201, 202]. For an alternative perspective, see reference [208].

13. The Effect of Neutrino Oscillations

Perhaps the most important discovery of particle physics in the last decade is the oscillation of neutrinos. The fact that neutrinos can change flavor as they propagate can have an effect on the neutrino fluxes observed on earth. For propagation over cosmological distances, the neutrino survival probablities become simple. For the example of neutrinos from pion decay, where there is an initial ratio of 1 to 2 to 0 for electron, muon and tau flavors, respectively, we expect approximately a ratio of 1 to 1 to 1 after oscillations. Therefore most of the event rates calculated for muon neutrinos have been overestimated by a factor of two. At sufficient energies, however, electron and tau neutrinos can be observed as well thus counteracting this effect in part.

For a review of neutrino oscillations, see Ref. [209].

C. Blazars: the Sources of the Highest Energy Gamma rays

1. Blazar Characteristics

Active Galactic Nuclei (AGN) are the brightest sources in the universe. They are of special interest here because some emit most of their luminosity at GeV energy and above. A subset, called blazars, emit high-energy radiation in collimated jets pointing at the earth. They have the following characteristics:

- Although less luminous than GRB, with inferred isotropic luminosities of $\sim 10^{45} - 10^{49}$ ergs, they radiate this luminosity over much longer time periods with regular flares extending for days [210, 211, 212]. The energetics require a black hole roughly one billion times more massive than our sun.

- Blazars produce radiation from radio waves to TeV gamma rays with enhancements in $E^2 dN/dE$ or $\nu F(\nu)$ in the IR to X-ray and the MeV-TeV range [213, 214, 215, 216, 217]. Roughly 60 sources have been observed in the MeV-GeV range by the EGRET instrument on the Compton Gamma ray Observatory. A handful of TeV observations have been reported thus far [24, 210, 218, 219, 220, 221, 222]. EGRET has found that blazars produce a typical $\phi_\gamma \propto E^{-2.2}$ spectrum in the MeV-GeV range. This spectrum may extend above or below this range [210, 211, 223]. There appears to be an inverse relationship between blazar luminosity and peak emission. High luminosity blazars tend to peak in the GeV and optical bands [211, 215, 224, 225], while low luminosity blazars tend to peak in the TeV and X-ray bands [211, 213, 225, 226, 227, 228].

- Blazars are cosmological sources. They appear less distant than GRB only because their lower luminosity makes distant sources difficult to observe [210, 211].

- The time scale of flaring in blazar luminosity varies from a fraction of a day to years. This range of time scales indicates a sub-parsec engine; $c\Delta t \sim R' \rightarrow R' \sim 10^{-3} - 10^{-1}$ pc [210, 211, 223, 229].

Prior to the launch of EGRET, the only survey above 100 MeV energy had been made by the COS-B satellite. The COS-B survey revealed the first extragalactic gamma ray source, the active galaxy 3C-272 [230]. After EGRET was launched in 1991, several gamma ray blazars were discovered. The majority of these were flat-spectrum radio-loud quasars, although some were BL Lac objects which emit into the TeV range [223, 229, 231].

Three of the EGRET sources have also been observed by ground-based atmospheric Cerenkov telescopes previously discussed. Future observations of blazars will include lower-threshold (below 50 GeV) ground-based telescopes, as well as next generation satellites such as GLAST [210, 212, 232].

2. Blazar Models

It is widely believed that blazars are powered by accreting supermassive black holes with masses of $\sim 10^7 M_\odot$ or more. Some of the infalling matter is reemitted and accelerated in highly beamed jets aligned with the rotation axis of the black hole.

It is generally agreed upon that synchrotron radiation by accelerated electrons is the source of the observed IR to X-ray peak in the spectrum [211, 231, 233, 234, 235]. Inverse Compton scattering of synchrotron or, possibly, other ambient photons by the same electrons can accommodate all observations of the MeV-GeV second peak in the spectrum. There is a competing explanation for the second peak, however. In hadronic models [211, 236, 237, 238, 239, 240, 241, 242], MeV-GeV gamma rays are generated by accelerated protons interacting with gas or radiation surrounding the black hole. Pions produced in these interactions decay into the observed gamma rays and not-yet observed neutrinos. This process is accompanied by synchrotron radiation of the protons.

The basic dynamics of the blazar is common to all models. Relativistic jets are generated with substructure that takes the form of "blobs" or "sheets" of matter traveling along the jet with Lorentz factors of 10-100. As previously discussed, in order to accommodate the observation of flares, the thickness of these sheets must be less than $\gamma c \Delta t \sim 10^{-2}$ parsec, much smaller than their width, which is of the order of 1 parsec. It is in these blobs that shocks produce TeV gamma rays and high-energy neutrinos.

Several calculations of the neutrino flux from active galactic nuclei have been performed [137, 243, 244, 245, 246, 247, 248, 249, 250, 251, 252, 253, 254]. In the following sections, we describe two examples that illustrate the mechanisms for neutrino production in hadronic blazars.

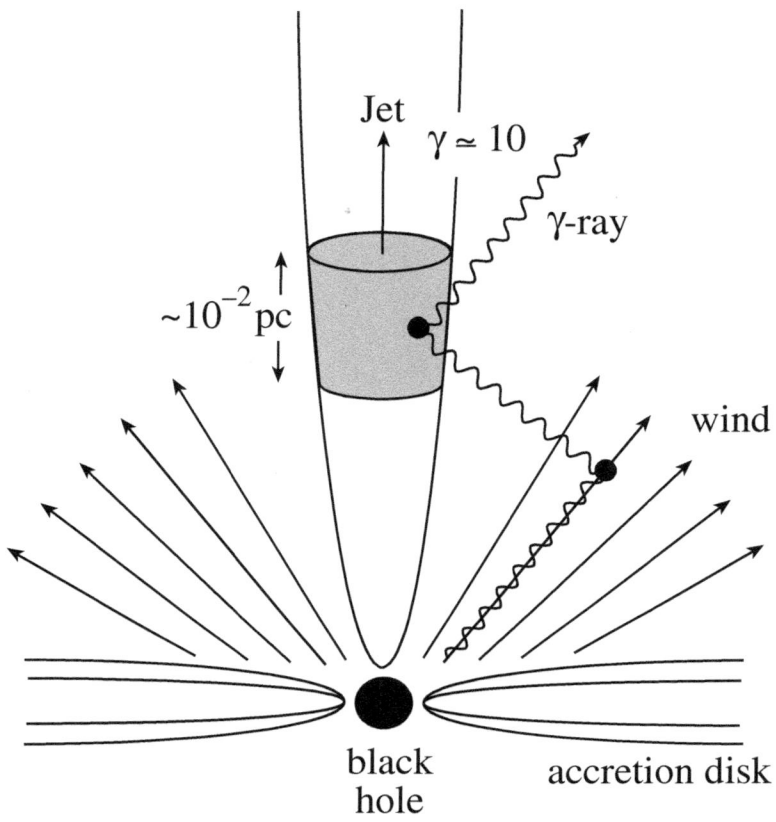

FIG. 18: Diagram of Blazar Kinematics

3. Highly Shocked Protons: EeV Blazar Neutrinos

If protons are present in blazar jets, they may interact with photons via the Δ-resonance to generate pions which then decay into very high-energy neutrinos. This process is similar to the process generating PeV and EeV neutrinos in GRB. There are some important differences, however. First, the Lorentz factor of the motion of the blob, traveling towards the observer, is typically smaller than for GRB shells. It can be constrained by considering the energy carried by the highest energy gamma rays observed in blazars. These gamma rays escape the blob and must, therefore, be below the energy threshold for pair production with ambient photons whose energy typically peaks around ~ 10 eV; the UV bump. Contemplating the observations of gamma rays above 15 TeV in Markarian 501 [218, 219], evading pair production requires

$$E_{\gamma,\text{max}} E_{\gamma,\text{peak}} < \gamma^2 m_e^2, \tag{94}$$

$$\gamma > 25 \left(\frac{E_{\gamma,\text{max}}}{15\,\text{TeV}}\right)^{1/2} \left(\frac{E_{\gamma,\text{peak}}}{10\,\text{eV}}\right)^{1/2}. \tag{95}$$

We will therefore consider Lorentz factors in the range of 10 to 100.

A second difference between blazars and GRB is the geometry of the shocked material. Instead of a shell, blobs can be treated as roughly spherical. The energy density is

$$U'_\gamma = \frac{L'_\gamma \Delta t}{\frac{4}{3}\pi R'^3} = \frac{L_\gamma \Delta t}{\gamma \frac{4}{3}\pi(\gamma c \Delta t)^3} = \frac{3 L_\gamma}{4\pi c^3 \gamma^4 \Delta t^2} \tag{96}$$

Except for geometry, this is identical to Eq. 49. We obtain a number density of photons

$$n_\gamma = \frac{U'_\gamma}{E'_\gamma} = \frac{3 L_\gamma}{4 E_\gamma \pi c^3 \gamma^3 \Delta t^2} \tag{97}$$

Following the arguments leading to Eqs. 51, we obtain a rather large conversion of energy into pions

$$f_\pi \simeq \frac{R'}{\lambda_{p\gamma}} \simeq \frac{L_\gamma}{E_\gamma} \frac{1}{\gamma^2 \Delta t} \frac{3\sigma_\Delta \langle x_{p\to\pi}\rangle}{4\pi c^2} \simeq .35 \times \left(\frac{L_\gamma}{10^{45}\text{erg/s}}\right) \left(\frac{10\,\text{eV}}{E_\gamma}\right) \left(\frac{30}{\gamma}\right)^2 \left(\frac{1000\,\text{sec}}{\Delta t}\right) \tag{98}$$

When f_π approaches unity, pions will be absorbed before decaying into neutrinos, requiring the substitution of f_π by $1 - e^{-f_\pi}$. Note that fluctuations in γ are not as important as for GRB because the fraction of energy transferred to pions varies as γ^2 rather than γ^4. Moreover, it quickly saturates near unity, especially for relatively low γ, i.e. bright flares or short time scales.

For protons to photoproduce pions on photons with the ubiquitous UV photons of $\sim 10\,\text{eV}$ energy,

$$E'_p > \frac{m_\Delta^2 - m_p^2}{4 E'_\gamma}. \tag{99}$$

Therefore, in the observer's frame,

$$E_p > 1.4 \times 10^{19}\text{eV} \left(\frac{\gamma}{30}\right)^2 \left(\frac{10\,\text{eV}}{E_\gamma}\right). \tag{100}$$

If blazars are the sources of the highest energy cosmic rays, protons are accelerated to this energy and will generate accompanying neutrinos with energy

$$E_\nu = \frac{1}{4}\langle x_{p\to\pi}\rangle E_p > 7 \times 10^{17}\text{eV} \left(\frac{\gamma}{30}\right)^2 \left(\frac{10\,\text{eV}}{E_\gamma}\right). \tag{101}$$

The neutrino flux from blazars can be calculated in the same way as for GRB

$$\phi_\nu \simeq \frac{c}{4\pi} \frac{1}{E_\nu} \left(\frac{1}{2}(1-e^{-f_\pi})t_H \frac{dE}{dt}\right) e^{(1-e^{-f_\pi})}, \tag{102}$$

$$\phi_\nu = 10^{-15} \text{cm}^{-2}\text{s}^{-1} \left(\frac{7 \times 10^{17}\text{eV}}{E_\nu}\right) \left(\frac{t_H}{10\text{Gyr}}\right) \left(\frac{dE/dt}{4 \times 10^{44}}\right) (1-e^{-f_\pi}) e^{(1-e^{-f_\pi})}, \tag{103}$$

using Eq. 52 and Eq. 53. This a flux on the order of 100 km^{-2}yr^{-1} over 2π steridian. The number of detected events is obtained from Eqs. 11,

$$N_{\text{events}} \sim \phi_\nu P_{\nu \to \mu} \sim 10 \text{ km}^{-2}\text{yr}^{-1} \left(\frac{7 \times 10^{17}\text{eV}}{E_\nu}\right)^{1/2} \left(\frac{t_H}{10\text{Gyr}}\right) \left(\frac{dE/dt}{4 \times 10^{44}\text{Mpc}^{-3}\text{yr}^{-1}}\right) \tag{104}$$

for $f_\pi \simeq .35$. For values of γ varying from 10 to 100, the number of events varies from 1 to 70 events km^{-2}yr^{-1}, respectively. Observation in a kilometer-scale detector should be possible.

The greatest uncertainty in this calculation is associated with the requirement that blazars accelerate protons to the highest observed energies. In the next section we present an alternative mechanism for producing neutrinos in blazars which does not invoke protons of such high-energy.

4. Moderately Shocked Protons: TeV Blazar Neutrinos

In line-emitting blazars, external photons with energies in the keV-MeV energy range are known to exist. They are clustered in clouds of quasi-isotropic radiation. Protons of lower energy relative to those contemplated in the previous section, can photoproduce pions in interactions with these clouds [250, 251, 252, 253]. Consider a target of external photons of energy near $E_{\gamma,\text{ext}}$ with a luminosity $L_{\gamma,\text{ext}}$. The fraction of proton energy transfered to pions is approximately given by

$$f_{\pi,\text{ext}} \simeq f_{\pi,\text{int}} \frac{L_{\gamma,\text{ext}}}{L_{\gamma,\text{int}}} \frac{E_{\gamma,\text{ext}}}{E_{\gamma,\text{int}}}. \tag{105}$$

The neutrino energy threshold is

$$E_\nu > 7 \times 10^{13} \text{eV} \left(\frac{\gamma}{30}\right)^2 \left(\frac{100 \text{ keV}}{E_{\gamma,\text{ext}}}\right). \tag{106}$$

We will no longer relate the flux of protons to cosmic rays. Instead we introduce the luminosity of protons, L_p above pion production threshold. This is a largely unknown parameter although it has been estimated to be on the order of 10% of the total luminosity [249]. The proton energy needed to exceed the threshold of Eq. 44 is

$$E_p > 1.4 \times 10^{15} \left(\frac{\gamma}{30}\right)^2 \left(\frac{100 \text{ keV}}{E_{\gamma,\text{ext}}}\right). \tag{107}$$

The neutrino flux can be calculated as a function of L_p

$$\Phi_\nu \simeq \frac{\frac{1}{2}\langle x_{p\to\pi}\rangle L_p f_{\pi,\text{ext}} \Delta t}{E_\nu 4\pi D^2}, \tag{108}$$

where Δt is the duration of a blazar flare. This reduces to

$$\Phi_\nu \sim 4 \times 10^4 \text{ km}^{-2} \left(\frac{f_{\pi,\text{ext}}}{.5}\right)\left(\frac{L_p}{10^{45} \text{ erg/s}}\right)\left(\frac{\Delta t}{1000 \text{ sec}}\right)\left(\frac{1000 \text{ Mpc}}{D}\right)^2 \left(\frac{30}{\gamma}\right)^2 \left(\frac{E_{\gamma,\text{ext}}}{100 \text{ keV}}\right) \tag{109}$$

for a fifteen minute flare. Using Eqs. 11, the event rate of TeV neutrinos is

$$N_{\text{events}} \sim \phi_\nu P_{\nu\to\mu} \sim 2 \text{ km}^{-2}\left(\frac{f_{\pi,\text{ext}}}{.5}\right)\left(\frac{L_p}{10^{45} \text{ erg/s}}\right)\left(\frac{\Delta t}{1000 \text{ sec}}\right)\left(\frac{1000 \text{ Mpc}}{D}\right)^2 \left(\frac{30}{\gamma}\right)^{2/5} \left(\frac{E_{\gamma,\text{ext}}}{100 \text{ keV}}\right)^{1/5} \tag{110}$$

Note that this result is for a typical, but fairly distant source. A nearby line-emitting blazar could be a strong candidate for neutrino observation. Considering the more than 60 blazars which have been observed, the total flux may generate conservatively tens or, optimistically, hundreds of TeV-PeV neutrino events per year in a kilometer scale neutrino telescope such as IceCube. The upper range of this estimate can be explored by the AMANDA experiment.

Blazar neutrino searches should be able to find incontrovertible evidence for cosmic ray acceleration in active galaxies, or, alternatively, challenge the possibility that AGN are the sources of the highest energy cosmic rays.

D. Neutrinos Associated With Cosmic Rays of Top-Down Origin

In addition to astrophysical objects such as GRB and blazars, a variety of top-down models have been proposed as the source of the highest energy cosmic rays. For example, annihilating or decaying superheavy relic particles could produce the highest energy cosmic rays [126, 130, 255, 256, 257, 258, 259, 260, 261, 262]. In addition to the cosmic ray nucleons, they will also generate gamma-rays and neutrinos. Other top-down scenarios which solve the cosmic ray problem in a similar way include topological defects [131, 132, 263, 264, 265, 266] and Z-bursts [114, 115, 116, 117, 118]. Conventional particle physics implies that ultra high-energy jets fragment predominantly into photons with a small admixture of protons

[267, 268, 269]. This seems to be in disagreement with mounting evidence that the highest energy cosmic rays are not photons [4]. In light of this information, we must assume that protons, and not gamma-rays, dominate the highest energy cosmic ray spectrum. This does not necessarily rule out superheavy particles as the source of the highest energy cosmic rays. The uncertainties associated with the universal radio background and the strength of intergalactic magnetic fields leave open the possibility that ultra high-energy photons may be depleted from the cosmic ray spectrum near 10^{20} eV, leaving a dominant proton component at GZK energies [270, 271, 272, 273]. With this in mind, one must normalize the proton spectrum from top-down scenarios with the observed ultra high-energy cosmic ray flux.

An important point is that this "renormalization" is not only challenged by the large sub-GeV photon flux, but also by the neutrino flux associated with these models. Neutrinos, which are produced more numerously than protons and travel much greater distances, typically provide an observable signal in operating high-energy neutrino telescopes.

1. Nucleons in Top-Down Scenarios

The assumption that nucleons from ultra high-energy fragmentation are the source of the highest energy cosmic rays normalizes the rate of their generation. To do this, it is necessary to calculate the spectrum of nucleons produced in such jets. Each jet will fragment into a large number of hadrons. The quark fragmentation function can be parametrized as [274]:

$$\frac{dN_{\rm H}}{dx} = Cx^{-3/2}(1-x)^2. \tag{111}$$

Here, $x = E_{\rm Hadron}/E_{\rm Jet}$, $N_{\rm H}$ is the number of hadrons and $C = 15/16$ is a normalization constant determined by energy conservation. For a more rigorous treatment of fragmentation, see [127, 128, 129].

At the energies considered, all flavors of quarks are produced equally. Top quarks immediately decay into $bW^{\pm}$ pairs. Bottom and charm quarks lose energy from hadronization before decaying into charmed and other hadrons. Hadrons eventually decay into pions and nucleons. The injection spectrum of nucleons produced in parton jets can be approximately described by [275]:

$$\Phi_N(E) \simeq \frac{dn_X}{dt} N_q^2 \frac{f_N}{E_{\rm Jet}} \frac{dN_{\rm H}}{dx}. \tag{112}$$

Here, $\frac{dn_X}{dt}$ is the number of jets produced per second per cubic meter. N_q is the number of quarks produced per jet in the energy range concerned. $f_N \sim .03$ is the nucleon fraction in the jet from a single quark [275].

To solve the ultra high-energy cosmic ray problem, this proton flux must accommodate the events above the GZK cutoff. Observations indicate on the order of 10^{-27} events $m^{-2}s^{-1}sr^{-1}GeV^{-1}$ in the energy range above the GZK cutoff (5×10^{19} eV to 2×10^{20} eV) [1, 5]. The formalism of a generic top-down scenario is sufficiently flexible to explain the data from either the HIRES or AGASA experiments. The distribution of ultra high-energy jets can play an important role in the spectra of nucleons near the GZK cutoff. For example, the distribution for decaying or annihilating dark matter is likely to be dominated by the dark matter within our galaxy. This overdensity strongly degrades the effect of the GZK cutoff.

2. Neutrinos in Top-Down Scenarios

There are several ways neutrinos can be produced in the fragmentation of ultra high-energy jets. First bottom and charm quarks decay semileptonically about 10% of the time. Secondly, the cascades of hadrons produce mostly pions. About two thirds of these pions will be charged and decay into neutrinos [275]. Furthermore, top quarks produced in the jets decay nearly 100% of the time to $bW^\pm$. The W bosons then decay semileptonically approximately 10% of the time to each neutrino species.

Generally, the greatest contribution to the neutrino spectrum is from charged pions. The injection spectrum of charged pions is given by [275]:

$$\Phi_{(\pi^+ + \pi^-)}(E) \simeq \frac{2}{3}\frac{(1-f_N)}{f_N}\Phi_N(E) \tag{113}$$

$$\Phi_{(\pi^+ + \pi^-)}(E) \simeq \frac{2}{3}(1-f_N)\frac{dn_X}{dt}\frac{N_q^2}{E_{\text{Jet}}}\frac{dN_{\text{H}}}{dx}. \tag{114}$$

The resulting injection of neutrinos is given by [276]:

$$\Phi_{(\nu+\bar{\nu})}(E) \simeq 2.34 \int_{2.34E}^{E_{\text{Jet}}/N_q} \frac{dE_\pi}{E_\pi}\Phi_{(\pi^++\pi^-)}(E_\pi). \tag{115}$$

Using $f_N \simeq .03$ and $\frac{dn_X}{dt} \simeq 1.5 \times 10^{-37}$, this becomes:

$$\Phi_{(\nu+\bar{\nu})}(E) \sim 3.0 \times 10^{-36} \int_{2.34E}^{E_{\text{Jet}}/N_q} \frac{dE_\pi}{E_\pi} N_q^2 (1-\frac{E_\pi}{E_{\text{Jet}}})^2 \frac{E_{\text{Jet}}^{.5}}{E_\pi^{1.5}} \tag{116}$$

for each species of neutrino. N_q is the number of quarks produced in the fragmentation in the energy range of interest.

To obtain the neutrino flux, we multiply the injection spectrum by the average distance traveled by a neutrino and by the rate per volume for hadronic jets which we calculated earlier. Neutrinos, not being limited by scattering, travel up to the age of the universe at the speed of light ($\sim$ 3000 Mpc in an Euclidean approximation). A random cosmological distribution of ultra high-energy jets provides an average distance between 2000 and 2500 Mpc.

The neutrinos generated in these scenarios can be constrained by measurements of the high-energy diffuse flux. AMANDA-B10, with an effective area of $\sim$5,000 square meters has placed the strongest limits on the flux at this time. In addition to the diffuse flux of high-energy neutrinos, the number of extremely high-energy events can be considered. Depending on the details of fragmention and jet distribution, tens to thousands of events per year per square kilometer effective area can be generated above an energy threshold of 1 PeV where there are no significant backgrounds to interfere with the signal.

As a simple example, take the Z-burst scenario. In this scenario, ultra high-energy neutrinos travel cosmological distances and interact with massive ($\sim$eV) cosmic background neutrinos at the Z-resonance. The Z bosons then decay producing, among other things, the super-GZK cosmic rays. In the center-of-mass frame of the neutrino annihilation the Z is produced at rest with all the features of its decay experimentally known. Independent of any differences in the calculation, normalizing the cosmic ray flux to the protons, rather than the photon flux, raises the sensitivity of neutrino experiments as in all other examples. This can be demonstrated with a simple calculation. Data determine that Z-decays produce 8.7 charged pions for every proton and, therefore, $8.7 \times 3 = 26.1$ neutrinos from $\pi^{\pm} \rightarrow \bar{\nu}_\mu \mu \rightarrow e \nu_\mu \bar{\nu}_e$ for every proton. AGASA data, with an integrated proton flux of $\sim 5 \times 10^{24}\,\mathrm{ev^2\,m^{-2}\,s^{-1}\,sr^{-1}}$ in the range of, say, 3×10^{19} to 2×10^{20}eV, indicate $\sim$ .5 protons per square kilometer, per year over 2π steridians. This normalization, corrected for the fact that neutrinos travel cosmological distances rather than a GZK radius for protons, predicts $.5 \times 26.1 \times \frac{3000\,\mathrm{Mpc}}{50\,\mathrm{Mpc}} \sim 800$ neutrinos per square kilometer, per year over 2π steridians. We here assumed an isotropic distribution of cosmological sources. The probability of detecting a neutrino at $\sim$ 10 EeV is $\sim$ .05, see Eq. 11. Therefore, we expect 40 events per year in IceCube, or a few events per year in AMANDA II. As with other top-down scenarios, present experiments are near excluding or confirming the model.

Using the high-energy neutrino diffuse flux measurements and searches for super-PeV neutrinos, top-down scenarios can be constrained. Further data from AMANDA, or next generation neutrino telescope IceCube, will test the viability of top-down scenarios which generate the highest energy cosmic rays.

IV. THE FUTURE FOR HIGH-ENERGY NEUTRINO ASTRONOMY

At this time, neutrino astronomy is in its infancy. Two telescopes, one in Lake Baikal and another embedded in the South Pole glacier, represent proof of concept that natural water and ice can be transformed into large volume Cherenkov detectors. With an acceptance of order $0.1\ km^2$, the operating AMANDA II telescope represents a first-generation instrument with the potential to detect neutrinos from sources beyond the earth's atmosphere and the sun. It has been operating for 3 years with 302 OM and for almost 3 years with 677 OM. Only 1997 data have been published. While looking forward to AMANDA data, construction has started on ANTARES, NESTOR and IceCube, with first deployments anticipated in 2002 and 2003. At super-EeV energies these experiments will be joined by HiRes, Auger and RICE. A variety of novel ideas exploiting acoustic and radio detection techniques are under investigation, including ANITA for which a proposal has been submitted. Finally, initial funding of the R&D efforts towards the construction of a kilometer-scale telescope in the Mediterranean has been awarded to the NEMO collaboration. With the pioneering papers published nearly half a century ago by Greisen, Reines and Markov, the technology is finally in place for neutrino astronomy to become a reality. The neutrino, a particle that is almost nothing, may tell us a great deal about the universe [277].

Acknowledgments

This work was supported in part by a DOE grant No. DE-FG02-95ER40896 and in part by the Wisconsin Alumni Research Foundation.

---

[1] D. J. Bird *et al.*, *Phys. Rev. Lett.* **71**, 3401 (1993).

[2] N. N. Efimov et al., *ICRR Symposium on Astrophysical Aspects of the Most Energetic Cosmic Rays*, ed. M. Nagano and F. Takahara (World Scientific, 1991).

[3] http://www.hep.net/experiments/all_sites.html, provides information on experiments discussed in this review. For a few exceptions, We will give separate references to articles or websites.

[4] M. Ave et al., *Phys. Rev. Lett.* **85**, 2244 (2000).

[5] http://www-akeno.icrr.u-tokyo.ac.jp/AGASA/

[6] Proceedings of the International Cosmic Ray Conference, Hamburg, Germany, August 2001.

[7] R. A. Vazquez et al., *Astroparticle Physics* **3**, 151 (1995).

[8] K. Greisen, *Ann. Rev. Nucl. Science*, **10**, 63 (1960).

[9] F. Reines, *Ann. Rev. Nucl. Science*, **10**, 1 (1960).

[10] M.A. Markov & I.M. Zheleznykh, *Nucl. Phys.* **27** 385 (1961).

[11] M. A. Markov in *Proceedings of the 1960 Annual International Conference on High-energy Physics at Rochester*, E. C. G. Sudarshan, J. H. Tinlot & A. C. Melissinos, Editors (1960).

[12] F. W. Stecker and M. H. Salamon, Astrophys. J. **512**, 521 (1992), astro-ph/9808110.

[13] J. Alvarez-Muniz, F. Halzen, T. Han and D. Hooper, Phys. Rev. Lett. **88**, 021301 (2002), hep-ph/0107057.

[14] R. Emparan, M. Masip and R. Rattazzi, Phys. Rev. D **65**, 064023 (2002), hep-ph/0109287.

[15] P. Jain, D. W. McKay, S. Panda and J. P. Ralston, Phys. Lett. B **484**, 267 (2000), hep-ph/0001031.

[16] A. Jain, P. Jain, D. W. McKay and J. P. Ralston, hep-ph/0011310.

[17] C. Tyler, A. V. Olinto and G. Sigl, Phys. Rev. D **63**, 055001 (2001), hep-ph/0002257.

[18] S. Nussinov and R. Shrock, Phys. Rev. D **59**, 105002 (1999), hep-ph/9811323.

[19] S. Nussinov and R. Shrock, Phys. Rev. D **64**, 047702 (2001), hep-ph/0103043.

[20] G. Domokos and S. Kovesi-Domokos, Phys. Rev. Lett. **82**, 1366 (1999), hep-ph/9812260.

[21] G. Domokos, S. Kovesi-Domokos and P. T. Mikulski, hep-ph/0006328.

[22] J. L. Feng and A. D. Shapere, Phys. Rev. Lett. **88**, 021303 (2002), hep-ph/0109106.

[23] A. Ringwald and H. Tu, Phys. Lett. B **525**, 135 (2002), hep-ph/0111042.

[24] T. C. Weekes, Status of VHE Astronomy c.2000, *Proceedings of the International Symposium on High-energy Gamma ray Astronomy*, Heidelberg, June 2000, astro-ph/0010431.

[25] R. A. Ong, *XIX International Symposium on Lepton and Photon Interactions at High Ener-*

gies, Stanford, August 1999, hep-ex/0003014.

[26] R. A. Ong, Phys.Rep. **305**, 93, (1998).

[27] C.M. Hoffman, C. Sinnis, P. Fleury and M. Punch, Rev. Mod. Phys. **71**, 897 (1999),

[28] T.C. Weekes, *Proceedings of DPF'99*, UCLA (1999).

[29] E. Lorenz, talk at TAUP99, Paris, France (1999).

[30] M. De Naurois et al., Astrophys. J. **566**, 343 (2002), astro-ph/0107301.

[31] J. Cortina for the MAGIC collaboration, *Proceedings of the Very High Energy Phenomena in the Universe*, Les Arcs, France, January 20–27, 2001, astro-ph/0103393.

[32] http://veritas.sao.arizona.edu/

[33] http://hegra1.mppmu.mpg.de

[34] http://www.igpp.lanl.gov/ASTmilagro.html

[35] F. Halzen, *The case for a kilometer-scale neutrino detector*, in Nuclear and Particle Astrophysics and Cosmology, Proceedings of Snowmass 94, R. Kolb and R. Peccei, eds.

[36] F. Halzen, *The Case for a Kilometer-Scale Neutrino Detector: 1996*, in Proc. of the Sixth International Symposium on Neutrino Telescopes, ed. by M. Baldo-Ceolin, Venice (1996).

[37] F. Halzen and E. Zas, Astrophys. J. **488**, 669 (1997), astro-ph/9702193.

[38] J. N. Bahcall and E. Waxman, Phys. Rev. D **64** (2001), hep-ph/9902383.

[39] E. Waxman and J. N. Bahcall, Phys. Rev. D **59** (1999), hep-ph/9807282.

[40] K. Mannheim, R. J. Protheroe and J. P. Rachen, Phys. Rev. D **63**, 023003 (2001), astro-ph/9812398.

[41] J. P. Rachen, R. J. Protheroe and K. Mannheim, *presented at 19th Texas Symposium on Relativistic Astrophysics: Texas in Paris*, Paris, France, 14-18 Dec 1998, astro-ph/9908031.

[42] T.K. Gaisser, R.J. Protheroe & Todor Stanev, Ap.J. 492 (1998) 219.

[43] L. O'C. Drury, F.A. Aharonian & H.J. Völk, A & A 287 (1994) 959.

[44] J.A. Esposito, S.D. Hunter, G. Kanbach & P. Sreekumar, Ap.J. 461 (1996) 820.

[45] W. Bednarek and R. J. Protheroe, Phys. Rev. Lett. **79**, 2616 (1997).

[46] T. K. Gaisser, F. Halzen and T. Stanev, Phys. Rept. **258**, 173 (1995) [Erratum-ibid. **271**, 355 (1995)], hep-ph/9410384.

[47] J.G. Learned and K. Mannheim, *Ann. Rev. Nucl. Part. Science* **50**, 679 (2000).

[48] Cross sections tabulated by R. Gandhi, C. Quigg, M. H. Reno and I. Sarcevic, Astropart. Phys. **5**, 81 (1996).

[49] J. Alvarez-Muniz and F. Halzen, *Detection of Tau Neutrinos in IceCube* (2001).

[50] H. Athar, G. Parente and E. Zas, Phys. Rev. D62 (2000) 093010, hep-ph/0006123.

[51] J. G. Learned and Sandip Pakvasa, *Astropart. Phys.* **3**, 267 (1995).

[52] IceCube Design Document, www.icecube.wisc.edu.

[53] F. Halzen and D. Hooper, *in preparation*.

[54] F. Halzen and D. Saltzberg, Phys. Rev. Lett. **81**, 4305 (1998).

[55] T. Stanev, Phys. Rev. Lett. **83**, 5427 (1999), astro-ph/9907018.

[56] J. F. Beacom, P. Crotty and E. W. Kolb, astro-ph/0111482.

[57] S. I. Dutta, M. H. Reno and I. Sarcevic, Phys. Rev. D **62**, 123001 (2000), hep-ph/0005310.

[58] F. Halzen, Nucl. Phys. Proc. Suppl. **100**, 320 (2001).

[59] Cosmic Rays in the Deep Ocean, the DUMAND Collaboration (J. Babson et al.). ICR-205-89-22, Dec 1989, 24pp, Published in Phys.Rev.D **42**, 3613 (1990).

[60] P. C. Bosetti [DUMAND Collaboration], *In Lohusalu 1995, Neutrino physics* 57-64.

[61] J. W. Bolesta et al., *Proc. 25th ICRC* Durban-South Africa, **7** (1997) 29.

[62] V. A. Balkanov et al, Nucl.Phys.Proc.Suppl. **75A**, 409 (1999).

[63] V. Balkanov et al. [BAIKAL Collaboration], *proceedings of 9th International Symposium on Neutrino Telescopes*, Venice, Italy, 6-9 Mar 2001, astro-ph/0105269.

[64] I. A. Belolaptikov et al., *Astroparticle Physics* **7**, 263 (1997).

[65] V. A. Balkanov et al. Astro. Part. Phys. **14**, 61 (2000).

[66] L. Trascatti, in *Procs. of the 5th International Workshop on Topics in Astroparticle and Underground Physics (TAUP 97)*, Gran Sasso, Italy, 1997, ed. by A. Bottino, A. di Credico, and P. Monacelli, Nucl. Phys. **B70** (Proc. Suppl.), p. 442 (1998).

[67] P. K. Grieder [NESTOR Collaboration], Nuovo Cim. **24C**, 771 (2001).

[68] L. Trasatti [NESTOR Collaboration], Nucl. Phys. Proc. Suppl. **70**, 442 (1999).

[69] E. Aslanides et al, astro-ph/9907432 (1999).

[70] F. Feinstein [ANTARES Collaboration], Nucl. Phys. Proc. Suppl. **70**, 445 (1999).

[71] T. Montaruli [ANTARES Collaboration], *for the proceedings of TAUP 2001: Topics in Astroparticle and Underground Physics*, Assergi, Italy, 8-12 Sep 2001, hep-ex/0201009.

[72] M. Spiro, *presentation to AASC Committee of the National Academy of Sciences*, Atlanta (1999).

[73] Talk given at the *International Workshop on Next Generation Nucleon Decay and Neutrino Detector (NNN 99)*, Stony Brook, 1999, Proceedings to be published by AIP.

[74] The AMANDA collaboration, *Astroparticle Physics*, **13**, 1 (2000).

[75] E. Andres *et al.*, *Nature* **410**, 441 (2001).

[76] R. Porrata *et al.*, *Proc. 25th ICRC* Durban-South Africa, **7** (1997) 9.

[77] M. Kowalski *et al.* [AMANDA Collaboration], *To appear in the proceedings of International Europhysics Conference on High-Energy Physics (HEP 2001)*, Budapest, Hungary, 12-18 Jul 2001, hep-ph/0112083.

[78] A. Karle, for the AMANDA collaboration, Observation of Atmospheric Neutrinos with the AMANDA Experiment, to be published in *Proceedings of the 17th International Workshop on Weak Interactions and Neutrinos*, Cape Town, South Africa (1999).

[79] P.Askjeber et al., *Science* **267**, 1147 (1995).

[80] G. C. Hill, Proceedings of the 26th International Cosmic Ray Conference, Salt Lake City, Utah (1999), astro-ph/0106081.

[81] G. C. Hill and M. J. Leuthld *et al.* [AMANDA Collaboration], Proceedings of the 27th International Cosmic Ray Conference, Hamburg (2001).

[82] F. Halzen, Talk given at *International Symposium on High Energy Gamma-Ray Astronomy*, Heidelberg, Germany, 26-30 Jun 2000, astro-ph/0103195.

[83] F. Halzen *et al.* [AMANDA Collaboration], *Prepared for 26th International Cosmic Ray Conference (ICRC 99), Salt Lake City, UT, 17-25 Aug 1999*.

[84] http://www.icecube.wisc.edu.

[85] J. Alvarez-Muniz and F. Halzen, AIP Conf. Proc. **579**, 305 (2001), astro-ph/0102106.

[86] J. Alvarez-Muniz and F. Halzen, Phys. Rev. D **63**, 037302 (2001), astro-ph/0007329.

[87] J. Alvarez-Muniz and E. Zas, AIP Conf. Proc. **579**, 128 (2001), astro-ph/0102173.

[88] D. Saltzberg *et al.*, Phys. Rev. Lett. **86**, 2802 (2001), hep-ex/0011001.

[89] E. Zas, F. Halzen and T. Stanev, Phys. Rev. D **45**, 362 (1992).

[90] I. Kravchenko *et al.* [RICE Collaboration], Submitted to Astropart.Phys., astro-ph/0112372.

[91] J. L. Feng, P. Fisher, F. Wilczek and T. M. Yu, Phys. Rev. Lett. **88**, 161102 (2002), hep-ph/0105067.

[92] P. Gorham, http://astro.uchicago.edu~olinto/aspen/program.

[93] L. G. Dedenko, I. M. Zheleznykh, S. K. Karaevsky, A. A. Mironovich, V. D. Svet and

A. V. Furduev, Bull. Russ. Acad. Sci. Phys. **61** (1997) 469 [Izv. Ross. Akad. Nauk. **61** (1997) 593].

[94] E. Waxman and J. N. Bahcall, Phys. Rev. Lett. **78**, 2292 (1997), astro-ph/9701231.

[95] M. Vietri, Phys. Rev. Lett. **80**, 3690 (1998), astro-ph/9802241.

[96] M. Bottcher and C. D. Dermer, Submitted to Astrophys.J.Lett., astro-ph/9801027.

[97] A. V. Olinto, R. I. Epstein and P. Blasi, Proceedings of the 26th International Cosmic Ray Conference, Salt Lake City, Utah, USA (1999).

[98] V. S. Berezinsky and O. F. Prilutsky *astron. Astrophys.* **66** (1978) 325.

[99] H. Sato, Prog. Theor. Phys. **58**, 549 (1977).

[100] T. K. Gaisser and T. Stanev, Phys. Rev. Lett. **54**, 2265 (1985).

[101] E. W. Kolb, M. S. Turner and T. P. Walker, Phys. Rev. D **32**, 1145 (1985) [Erratum-ibid. D **33**, 859 (1985)].

[102] V. S. Berezinsky, C. Castagnoli and P. Galeotti, Nuovo Cim. **8C**, 185 (1985) [Addendum-ibid. **8C**, 602 (1985)].

[103] A. Levinson and E. Waxman, Phys. Rev. Lett. **87**, 171101 (2001), hep-ph/0106102.

[104] C. Distefano, D. Guetta, E. Waxman and A. Levinson, Submitted to Astrophys.J., astro-ph/0202200.

[105] D. Seckel, T. Stanev and T. K. Gaisser, Astrophys. J. **382**, 652 (1991).

[106] C. Hettlage, K. Mannheim and J. G. Learned, Astropart. Phys. **13**, 45 (2000), astro-ph/9910208.

[107] F. W. Stecker, Astrophys. J. **228**, 919 (1979).

[108] R. Engel, D. Seckel and T. Stanev, Phys. Rev. D **64**, 093010 (2001), astro-ph/0101216, and references therein.

[109] C. G. Costa, Astropart. Phys. **16**, 193 (2001), hep-ph/0010306, and references therein.

[110] C. G. Costa, F. Halzen and C. Salles, hep-ph/0104039.

[111] V. S. Berezinsky, T. K. Gaisser, F. Halzen and T. Stanev, Astropart. Phys. **1**, 281 (1993).

[112] V. S. Berezinsky and V. A. Kudryavtsev, *Sov. Astron. Lett.* **14** 873 (1998).

[113] V. S. Berezinsky, *Prepared for International Workshop on Neutrino Telescopes, 3rd, Venice, Italy, 26-28 Feb 1991.*

[114] T. Weiler, Phys. Rev. Lett. **49**, 234 (1982).

[115] T. J. Weiler, Astropart. Phys. **11**, 303 (1999), hep-ph/9710431.

[116] D. Fargion, B. Mele and A. Salis, Astrophys. J. **517**, 725 (1999), astro-ph/9710029.

[117] Z. Fodor, S. D. Katz and A. Ringwald, Phys. Rev. Lett. **88**, 171101 (2002), hep-ph/0105064.

[118] Z. Fodor, S. D. Katz and A. Ringwald, hep-ph/0203198.

[119] V. D. Barger, F. Halzen, D. Hooper and C. Kao, Phys. Rev. D **65**, 075022 (2002), hep-ph/0105182.

[120] G. Eigen, R. Gaitskell, G. D. Kribs and K. T. Matchev, in *Proc. of the APS/DPF/DPB Summer Study on the Future of Particle Physics (Snowmass 2001)* ed. R. Davidson and C. Quigg, hep-ph/0112312.

[121] J. L. Feng, K. T. Matchev and F. Wilczek, in *Proc. of the APS/DPF/DPB Summer Study on the Future of Particle Physics (Snowmass 2001)* ed. R. Davidson and C. Quigg, hep-ph/0111295.

[122] J. L. Feng, K. T. Matchev and F. Wilczek, Phys. Rev. D **63**, 045024 (2001), astro-ph/0008115.

[123] J. R. Ellis, T. Falk, G. Ganis and K. A. Olive, Phys. Rev. D **62**, 075010 (2000), hep-ph/0004169.

[124] L. Bergstrom, J. Edsjoe and P. Gondolo, *presented at 26th International Cosmic Ray Conference (ICRC 99)*, Salt Lake City, UT, 17-25 Aug 1999. In Salt Lake City 1999, Cosmic ray, vol. 2 281-284, astro-ph/9906033.

[125] P. Gondolo and J. Silk, Phys. Rev. Lett. **83**, 1719 (1999), astro-ph/9906391.

[126] F. Halzen and D. Hooper, *proceedings of Dark Matter 2002*, Los Angeles, CA (2002), hep-ph/0110201.

[127] S. Sarkar and R. Toldra, Nucl. Phys. B **621**, 495 (2002), hep-ph/0108098.

[128] M. Birkel and S. Sarkar, Astropart. Phys. **9**, 297 (1998), hep-ph/9804285.

[129] C. Barbot and M. Drees, hep-ph/0202072.

[130] V. Berezinsky, M. Kachelriess and A. Vilenkin, Phys. Rev. Lett. **79**, 4302 (1997), astro-ph/9708217.

[131] V. S. Berezinsky and A. Vilenkin, Phys. Rev. D **62**, 083512 (2000), hep-ph/9908257.

[132] R. Gandhi, C. Quigg, M. H. Reno and I. Sarcevic, *Presented at 1996 Annual Divisional Meeting (DPF 96) of the Division of Particles and Fields of the American Physical Society*, Minneapolis, MN, 10-15 Aug 1996, hep-ph/9609516.

[133] F. Halzen, B. Keszthelyi and E. Zas, Phys. Rev. D **52**, 3239 (1995), hep-ph/9502268.

[134] E. V. Bugaev and K. V. Konishchev, astro-ph/0103265.

[135] S. W. Hawking, Nature **248**, 30 (1974).

[136] T. Piran, Phys. Rept. **314** (1999) 575, astro-ph/9810256.

[137] F. Halzen, *Lectures given at Theoretical Advanced Study Institute in Elementary Particle Physics (TASI 98): Neutrinos in Physics and Astrophysics: From 10^{-33} to 10^{+28} cm*, Boulder, CO, 31 May - 26 Jun 1998. Published in Boulder 1998, Neutrinos in physics and astrophysics 524-569, astro-ph/9810368.

[138] M. Metzger, *et al.*, Nature **387**, 878 (1997).

[139] S. Kulkarni, *et al.*, in *Gamma Ray Bursts*, Proc. 5th Huntsville Symp (AIP:New York), astro-ph/0002168.

[140] E. Costa, *et al.*, Nature **387**, 783 (1997).

[141] J. Paradijs, *et al.*, Nature **386** 686 (1997).

[142] D. Frail, *et al.*, *Gamma Ray Bursts*, Proc. 5th Huntsville Symp (AIP:New York), astro-ph/9912171.

[143] C. Kouveliotou, *et al.*, Ap. J. Lett., **413**, L101 (1993).

[144] K. C. Walker, B. E. Schaefer and E. E. Fenimore, Submitted to Astrophys.J., astro-ph/9810271.

[145] http://gammaray.msfc.nasa.gov/batse/grb/lightcurve/

[146] R. W. Klebesadel, I. B. Strong, R. A. Olson, Ap. J. Lett., **182**, L85 (1973).

[147] T. Cline, Ap. J. Lett., **185**, L1 (1973).

[148] C. A. Meegan *et al.*, Nature **355**, 143 (1992).

[149] A. S. Fruchter *et al.*, astro-ph/9902236.

[150] V. V. Sokolov *et al.*, Submitted to Astron.Astrophys., astro-ph/9809111.

[151] J. U. Fynbo *et al.*, Submitted to Astron.Astrophys., astro-ph/0102158.

[152] A. MacFadyen and S. E. Woosley, Astrophys. J. **524**, 262 (1999), astro-ph/9810274.

[153] S. E. Woosley, Ap. J., **405**, 273 (1993).

[154] C. Thompson, MNRAS, 270, 480 (1994).

[155] V. Usov, MNRAS, 267, 1035 (1994).

[156] H. C. Spruit, A&A, 341, L1 (1999).

[157] M. Vietri, L. Stella, Ap. J. Lett., **507** L45 (1998).

[158] S. Inoue, D. Guetta, F. Pacini, astro-ph/0111591.

[159] B. Paczynski, Ap. J. Lett., **494**, L45 (1998), astro-ph/9706232.

[160] K. Abazajian, G. M. Fuller and X. Shi, *proceedings of International Conference on the Activity of Galaxies and Related Phenomena*, Byurakan, Armenia, 17-21 Aug 1998, astro-ph/9812287.

[161] B. Qin, X. P. Wu, M. C. Chu and L. Z. Fang, Submitted to Astrophys.J., astro-ph/9708095.

[162] A. Mitra, astro-ph/9803014.

[163] D. Eichler, M. Livio, T. Piran and D. N. Schramm, Nature **340**, 126 (1989).

[164] J. Goodman, A. Dar, S. Nussinov, Ap. J. Lett., **314**, L7 (1987).

[165] B. Paczynski, Acta Astronomica, **41**, 257 (1991).

[166] T. Piran, R. Narayan and A. Shemi, CFA-3352A.

[167] R. Narayan, T. Piran and A. Shemi, CFA-3263.

[168] B. Paczynski, Astrophys. J. **308**, L43 (1986).

[169] A. Mitra, Submitted to Mon.Not.Roy.Astron.Soc., astro-ph/0010311.

[170] I. Bombaci and B. Datta, Submitted to Astrophys.J., astro-ph/0001478.

[171] K. S. Cheng and Z. G. Dai, Phys. Rev. Lett. **77**, 1210 (1996), astro-ph/9510073.

[172] J. N. Reeves, *et al.*, Outflowing supernova ejecta measured in the X-ray afterglow of Gamma Ray Burst GRB 011211 (2002).

[173] L. S. Finn, S. D. Mohanty and J. D. Romano, Phys. Rev. D **60**, 121101 (1999) gr-qc/9903101.

[174] P. Meszaros, Science **291**, 79 (2001), astro-ph/0102255.

[175] R. A. Chevalier and Z. Y. Li, astro-ph/9904417.

[176] E. Waxman, *Lectures given at 11th International Symposium on Very-High-Energy Cosmic Ray Interactions: The Gleb Wataghin Centennial (ISVHECRI 2000)*, Campinas, Brazil, 17-21 Jul 2000, astro-ph/0103186.

[177] J. Goodman, Ap. J., **308** L47 (1986).

[178] B. Paczynski, Ap. J., **363** 218 (1990).

[179] A. Shemi, T. Piran, Ap. J. Lett., **365** L55 (1990).

[180] E. Waxman, Phys. Rev. Lett. **75**, 386 (1995), astro-ph/9505082.

[181] A. Dar, Submitted to Astrophys.J.Lett., astro-ph/9709231.

[182] A. Dar and A. De Rujula, astro-ph/0105094.

[183] A. Dar, AIP Conf. Proc. **565** (2001) 455, astro-ph/0101007.

[184] J. I. Katz, Ap. J., **422**, 248 (1994).

[185] P. Meszaros, P. Laguna and M. J. Rees, Astrophys. J. **415**, 181 (1993), astro-ph/9301007.

[186] B. R. Schaefer, Ap. J., **492**, 696 (1997).

[187] E. Cohen, et al., Ap. J., **480**, 330 (1997).

[188] G. B. Rybicki and A. P. Lightman, *Radiative Processes in Astrophysics* (1979).

[189] S. R. Kulkarni et al., Nature **398** (1999) 389, astro-ph/9902272.

[190] P. Meszaros, M. J. Rees and R. A. Wijers, New Astron. **4**, 303 (1999), astro-ph/9808106.

[191] J. E. Rhoads, Astrophys. J. **487**, L1 (1997), astro-ph/9705163.

[192] M. Milgrom, and V. Usov, Astrophy. J. **449**, L37 (1995);

[193] M. Vietri, Astrophys. J. **453**, 883 (1995).

[194] C. D. Dermer, Submitted to Astrophys.J., astro-ph/0005440.

[195] E. Waxman, Nucl. Phys. Proc. Suppl. **91**, 494 (2000), hep-ph/0009152.

[196] J. P. Rachen and P. Meszaros, AIP Conf. Proc. **428**, 776 (1997), astro-ph/9811266.

[197] J. Miralda-Escude and E. Waxman, Astrophys. J. **462**, L59 (1996), astro-ph/9601012.

[198] E. Waxman, Astrophys. J. **452**, L1 (1995), astro-ph/9508037.

[199] F. Halzen and G. Jaczko, Phys. Rev. D **54**, 2779 (1996).

[200] J. P. A. Clark and D. Eardley, Ap. J., **215**, 311 (1977).

[201] F. Halzen and D. W. Hooper, Astrophys. J. **527**, L93 (1999), astro-ph/9908138.

[202] J. Alvarez-Muniz, F. Halzen and D. W. Hooper, Phys. Rev. D **62**, 093015 (2000), astro-ph/0006027.

[203] P. Meszaros and E. Waxman, Phys. Rev. Lett. **87**, 171102 (2001).

[204] E. Waxman and J. N. Bahcall, Astrophys. J. **541**, 707 (2000), hep-ph/9909286.

[205] P. Meszaros and E. Waxman, Phys. Rev. Lett. **87**, 171102 (2001), astro-ph/0103275.

[206] J. N. Bahcall and P. Meszaros, Phys. Rev. Lett. **85**, 1362 (2000), hep-ph/0004019.

[207] P. Meszaros and M. J. Rees, Astrophys. J. **541**, L5 (2000), astro-ph/0007102.

[208] D. Guetta, M. Spada and E. Waxman, Astrophys. J. **559**, 101 (2001), astro-ph/0102487.

[209] M. C. Gonzalez-Garcia and Y. Nir,' arXiv:hep-ph/0202058, and references therein.

[210] R. Mukherjee, *Invited talk at International Symposium on High Energy Gamma-Ray Astronomy*, Heidelberg, Germany, 26-30 Jun 2000, astro-ph/0101301.

[211] M. Sikora and G. Madejski, *Prepared for International Symposium on High Energy Gamma-Ray Astronomy*, Heidelberg, Germany, 26-30 Jun 2000, astro-ph/0101382.

[212] M. Catanese and T. C. Weekes, astro-ph/9906501.

[213] T. Takahashi et al., Astrophys. J. **470**, L89 (1996).

[214] T. Takahashi et al., Astrophys. J. **543**, L124 (2000).

[215] A. E. Wehrle *et al.*, Astrophys. J. **497**, 178 (1998).

[216] R. Edelson, Astrophys. J. **401**, 516 (1992).

[217] R. C. Hartman *et al.*, Astrophys. J. Suppl. **123**, 79 (1999).

[218] B. L. Dingus and D. L. Bertsch, Presented at *6th Compton Symposium on Gamma Ray Astrophysics 2001*, Baltimore, Maryland, 4-6 Apr 2001, astro-ph/0107053.

[219] R. J. Protheroe, C. L. Bhat, P. Fleury, E. Lorenz, M. Teshima and T. C. Weekes, Based on several talks given at *the International Cosmic Ray Conference (ICRC 97)* Durban, South Africa, 28 Jul - 8 Aug 1997, astro-ph/9710118.

[220] J. Kataoka *et al.*, Astrophys. J. **514**, 138 (1999).

[221] S.D. Hunter et al., Ap.J. 481 (1997) 205.

[222] F. Aharonian for the HEGRA Collaboration, astro-ph/9903386.

[223] R. Mukherjee *et al.*, Astrophys. J. **490**, 116 (1997).

[224] S. Wagner *et al.*, Astrophys. J. **454**, L97 (1995).

[225] L. Maraschi and F. Tavecchio, astro-ph/0102295.

[226] D. J. Macomb *et al.*, Astrophys. J. **449**, L99 (1995).

[227] D. Petry *et al.*, Astrophys. J. **536**, 742 (2000).

[228] R. M. Sambruna *et al.*, Astrophys. J. **538**, 127 (2000).

[229] C. von Montigny, *et al.*, Astrophys. J. **440**, 525 (1995).

[230] W. Hermsen *et al.*, Nature **269**, 494 (1977).

[231] R. Mukherjee *et al.*, astro-ph/9901106.

[232] H. F. Sadrozinski, Nucl. Instrum. Meth. A **466**, 292 (2001).

[233] P. Sreekumar, D. L. Bertsch, R. C. Hartman, P. L. Nolan and D. J. Thompson, Astropart. Phys. **11**, 221 (1999), astro-ph/9901277.

[234] M. Bottcher, *Proc. of the GeV-TeV Astrophysics International meeting*, Snowbird, Utah.

[235] M. Bottcher, Astrophys. J. **21**, L515 (1999).

[236] K. Mannheim, Astron. Astrophys. **269**, 67 (1993), astro-ph/9302006.

[237] K. Mannheim, Space Sci. Rev. **75**, 331 (1996).

[238] L. Nellen, K. Mannheim and P. L. Biermann, Phys. Rev. D **47**, 5270 (1993), hep-ph/9211257.

[239] A. Mucke and R. J. Protheroe, Astropart. Phys. **15**, 121 (2001), astro-ph/0004052.

[240] J. P. Rachen, Invited talk at *6th GeV - TeV Gamma Ray Astrophysics Workshop: Toward a Major Atmospheric Cerenkov Telescope*, Snowbird, Utah, 13-16 Aug 1999, astro-ph/0003282.

[241] R. J. Protheroe and A. Mucke, astro-ph/0011154.

[242] F. A. Aharonian, New Astron. **5**, 377 (2000), astro-ph/0003159.

[243] F. Stecker, C. Done, M. Salamon, and P. Sommers, *Phys. Rev. Lett.* **66**, 2697 (1991); erratum *Phys. Rev. Lett.* **69**, 2738 (1992).

[244] V. Berezinsky, Nucl. Phys. Proc. Suppl., **28A**, 352 (1992).

[245] F. W. Stecker and M. H. Salamon, Space Sci. Rev. **75**, 341 (1996).

[246] A. P. Szabo and R. J. Protheroe, *Proc. Workshop High-energy Neutrino Astronomy*, World Scientific, pg. 24 (1992).

[247] A. P. Szabo and R. J. Protheroe, Nucl. Phys. Proc. Suppl., **35**, 287 (1994).

[248] P. L. Biermann, *Proc. Workshop High-energy Neutrino Astronomy*, World Scientific, pg. 86 (1992).

[249] V. S. Berezinsky, Nucl. Phys. Proc. Suppl. **38**, 363 (1995).

[250] A. Atoyan and C. D. Dermer, Phys. Rev. Lett. **87**, 221102 (2001), astro-ph/0108053.

[251] C. D. Dermer and A. Atoyan, To appear in the *proceedings of 27th International Cosmic Ray Conferences (ICRC 2001)*, Hamburg, Germany, 7-15 Aug 2001, astro-ph/0107200.

[252] A. Mucke and R. J. Protheroe, To appear in the *proceedings of 27th International Cosmic Ray Conferences (ICRC 2001)*, Hamburg, Germany, 7-15 Aug 2001, astro-ph/0105543.

[253] R. J. Protheroe, Presented at *IAU Colloquium 163, Accretion Phenomena Related Outflows*, astro-ph/9607165.

[254] C. Schuster, M. Pohl and R. Schlickeiser, Submitted to Astron.Astrophys, astro-ph/0111545.

[255] E. W. Kolb, D. J. Chung and A. Riotto, To be published in the *proceedings of 2nd International Conference on Dark Matter in Astro and Particle Physics (DARK98)*, Heidelberg, Germany, 20-25 July 1998, hep-ph/9810361.

[256] D. J. Chung, E. W. Kolb and A. Riotto, Phys. Rev. D **59**, 023501 (1999), hep-ph/9802238.

[257] P. Blasi, R. Dick and E. W. Kolb, astro-ph/0105232.

[258] D. J. Chung, P. Crotty, E. W. Kolb and A. Riotto, Phys. Rev. D **64**, 043503 (2001), hep-ph/0104100; A. Riotto, *In Tegernsee 1999, Beyond the desert 503-521*.

[259] K. Benakli, J. R. Ellis and D. V. Nanopoulos, Phys. Rev. D **59**, 047301 (1999), hep-ph/9803333.

[260] C. Coriano, A. E. Faraggi and M. Plumacher, Nucl. Phys. B **614**, 233 (2001), hep-ph/0107053.

[261] S. Chang, C. Coriano and A. E. Faraggi, Nucl. Phys. B **477**, 65 (1996), hep-ph/9605325.

[262] H. Ziaeepour, Astropart. Phys. **16**, 101 (2001), astro-ph/0001137.

[263] R. J. Protheroe and T. Stanev, Phys. Rev. Lett. **77**, 3708 (1996), astro-ph/9605036.

[264] G. Sigl, S. Lee, P. Bhattacharjee and S. Yoshida, Phys. Rev. D **59**, 043504 (1999), hep-ph/9809242.

[265] P. Bhattacharjee, C. T. Hill and D. N. Schramm, Phys. Rev. Lett. **69**, 567 (1992).

[266] V. S. Berezinsky, P. Blasi, A. Vilenkin, Phys. Rev. D., (1999).

[267] R. Akers *et al.* [OPAL Collaboration], Z. Phys. C **63** (1994) 181.

[268] P. Abreu *et al.* [DELPHI Collaboration], Nucl. Phys. B **444** (1995).

[269] R. Barate *et al.* [ALEPH Collaboration], Phys. Rept. **294**, 1 (1998).

[270] P. Bhattacharjee and G. Sigl, Phys. Rept. **327**, 109 (2000),astro-ph/9811011.

[271] G. Sigl, S. Lee, P. Bhattacharjee and S. Yoshida, Phys. Rev. D **59**, 043504 (1999), hep-ph/9809242.

[272] R. J. Protheroe and T. Stanev, Phys. Rev. Lett. **77**, 3708 (1996), [Erratum-ibid. **78**, 3420 (1996)], astro-ph/9605036.

[273] R. J. Protheroe and P. L. Biermann, Astropart. Phys. **6**, 45 (1996), [Erratum-ibid. **7**, 181 (1996)], astro-ph/9605119.

[274] C. T. Hill, Nucl. Phys. B **224**, 469 (1983).

[275] P. Bhattacharjee and G. Sigl, Phys. Rept. **327**, 109 (2000), astro-ph/9811011.

[276] G. Jungman and M. Kamionkowski, Phys. Rev. D **51**, 328 (1995), hep-ph/9407351.

[277] *Spaceship Neutrino*, Christine Sutton (Cambridge University Press, Cambridge, 1992).

www.ingramcontent.com/pod-product-compliance
Lightning Source LLC
Chambersburg PA
CBHW070120210526
45170CB00013B/830